THE WILEY BICENTENNIAL—KNOWLEDGE FOR GENERATIONS

*E*ach generation has its unique needs and aspirations. When Charles Wiley first opened his small printing shop in lower Manhattan in 1807, it was a generation of boundless potential searching for an identity. And we were there, helping to define a new American literary tradition. Over half a century later, in the midst of the Second Industrial Revolution, it was a generation focused on building the future. Once again, we were there, supplying the critical scientific, technical, and engineering knowledge that helped frame the world. Throughout the 20th Century, and into the new millennium, nations began to reach out beyond their own borders and a new international community was born. Wiley was there, expanding its operations around the world to enable a global exchange of ideas, opinions, and know-how.

For 200 years, Wiley has been an integral part of each generation's journey, enabling the flow of information and understanding necessary to meet their needs and fulfill their aspirations. Today, bold new technologies are changing the way we live and learn. Wiley will be there, providing you the must-have knowledge you need to imagine new worlds, new possibilities, and new opportunities.

Generations come and go, but you can always count on Wiley to provide you the knowledge you need, when and where you need it!

WILLIAM J. PESCE
PRESIDENT AND CHIEF EXECUTIVE OFFICER

PETER BOOTH WILEY
CHAIRMAN OF THE BOARD

Graphing Calculator Guide for the TI-84 / 83

Functions
Modeling
Change

A Preparation for Calculus
Third Edition

by

Eric Connally
Harvard University Extension

Deborah Hughes-Hallett
University of Arizona

Andrew M. Gleason
Harvard University

et al.

John Wiley & Sons, Inc.

Associate Editor	Shannon Corliss
Production Editor	Carmen Hernandez

Wiley 200th Anniversary logo designed by: Richard J. Pacifico

Cover Photo: ©Grant V. Faint/Iconica/Getty Images, Inc.

The photo used on page 139 is provided courtesy of Texas Instruments Incorporated.

To order books or for customer service please, call 1-800-CALL WILEY (225-5945).

This material is based upon work supported by the National Science Foundation under Grant No. DUE-9352905. Opinions expressed are those of the authors and not necessarily those of the Foundation.

ISBN 978-0-470-10558-0

Printed in the United States of America

10 9 8 7 6 5 4 3 2 1

Printed and bound by Bind-Rite Graphics / Robbinsville.

PREFACE

The purpose of this book is to apply the features of the TI-84/83 graphing calculators to understand precalculus. The examples are taken from the precalculus book, *Functions Modeling Change, 3rd Edition* by Connally, Hughes-Hallett, Gleason, et al. I would like to acknowledge and thank the Calculus Consortium for Higher Education (CCHE) and John Wiley & Sons, Inc. for permission to freely use the text examples.

To the student

Using a graphing calculator can be both fun and frustrating. A healthy attitude when you get frustrated is to step back and say, "Isn't that interesting that it doesn't work." Figuring out how things work can be fun. If you get too frustrated, then it is time to ask a friend or the instructor for help.

Chapter Zero (Getting Started,) gives you a basic orientation to make you comfortable with how your calculator works. The remaining chapters, One to Twelve, correspond to the Table of Contents of the text *Functions Modeling Change*.

If you have a specific calculator question, your *TI–Guidebook* provides a resource that explains each feature briefly, and usually gives an example. (If you do not have the *TI–Guidebook*, it can be downloaded from the TI website.) Remember that the *TI-Guidebook* is like a dictionary: there is no story line or context. My intent here is to introduce the calculator features that you need for precalculus, so features are explained in the context of a precalculus example. Some calculator features that are less important to precalculus are not mentioned at all. The mathematical content drives this presentation, not the calculator features.

I have included tips about such things as shortcuts, warnings, and related ideas. I hope you find them useful.

> *Tip:*   Don't use technology in place of thinking.

To the instructor

It has been my aim to write this manual so that you can assign students to read it and practice the examples independently. It can also be used for in-class presentations. This manual focuses on precalculus examples, not on the calculator features. It has been my experience that the problems of teaching with this technology are greatly reduced by requiring a single standard calculator for a course. The TI-84 and TI-83 calculators are in most cases functionally interchangeable. However, the convenience of a USB computer connection on the TI-84 makes it the first choice.

Will these materials meet the needs of all your students? Of course not. There will still be the zealous students who want to write extensive programs and the anxious students who want all the buttons pressed for them. These materials are aimed at the middle, giving enough guidance so that most students are able to work through an example without assistance, but not so specific as to create a mindless exercise in pressing keys in the right order.

> *Tip:*   I strongly recommend using an overhead-projected calculator as a tool in your classroom. In most cases the TI Volume Purchase Plan can provide you with this package for free. I also recommend the software *TI-Connect*. Visit http://education.ti.com for resources.

Thanks

I would like to thank Seattle University for its continued support of office and computer facilities. I would like to acknowledge the debt I owe to Ann Davidian for her exuberant eagle-eyed editing of the first edition. Thanks to the folks at Wiley who have been a great help in answering questions and making arrangements.

— Carl Swenson, *swenson@seattleu.edu*

TABLE OF CONTENTS

12 PARAMETRIC EQUATIONS AND CONIC SECTIONS (CONT.)

CHAPTER ZERO

GETTING STARTED

Calculators have developed from the fingers, to the abacus, to the slide rule, to the scientific calculator, and now we have the graphing calculator. This chapter is a gentle introduction to using the TI-84/83 graphing calculator and shows how to make some simple numerical calculations. If you have used a graphing calculator before, you may only need to skim this chapter. The *TI–Guidebook* should also be consulted if you are having difficulty getting started. In this book the references to TI calculator keys and menu choices are written in the TI Uni font. The TI font looks like this.

A note on different models: TI-83, TI-83 Plus and TI-84

The oldest model, TI-83, is relatively slow and does not support the APPS feature. This feature is the ability to download, store and use special application software. An improvement, the TI-83 Plus, allows APPS, but has an inconvenient connection to a computer. TI-84 models are the fastest, have the most memory for storing APPS, and use a standard USB interface connection with computers. The TI-84 is the calculator of choice. On all three models, the keyboard and screen are similar, but not exactly the same. For example, the 2nd key on the TI-83 is written in lower case, while on the TI-84 it is in uppercase, 2ND. The keypads also differ in color scheme and key shape. The close similarity of keypad, screen and features makes it possible to use this book with any of the three models.

Basic Calculation

The ON key

Study the keyboard and press the ON key in the lower left-hand corner. You should see a blinking rectangular cursor. If not, then you may need to set the screen contrast. Even if the cursor is showing, it is a good idea to know how to adjust the screen contrast.

Using the 2nd key to adjust the screen contrast

As you use the calculator, the battery wears down and it becomes necessary to adjust the screen. Also, you may need to adjust the screen contrast for different lighting environments. Press and release the 2nd key and then the up-arrow key in the upper right of the keypad. By repeating this sequence, the screen darkens. The screen is lightened by repeating this sequence but by using the down-arrow instead of the up-arrow key. A momentary value (between 0 and 9) flashes in the upper right corner of the screen telling you the battery status (9 is close to replacement time). If the setting is too low, the cursor does not show; and if it is too high, the screen is dark as night.

If you take a break and come back later, the cursor disappears for a different reason. The calculator goes to sleep; it turns itself off after a few minutes of no activity. Just press the ON key and it wakes up at the same place it turned off: no memory loss.

Tip: Sometimes "broken" calculators can be fixed by reinserting the batteries correctly.

Arithmetic calculations on the home screen

Use technology on known results before trying complex examples. Our first use of the calculator is to do arithmetic on the home screen. Many different screens are shown later, but this *home* screen is where you do calculations. A graphing calculator has a distinct calculation advantage over a scientific calculator because it shows multiple lines and has entry recall. The numeric keys and the operation keys are used for simple arithmetic calculations.

Type some calculation for which you know the answer, say 8*9, and press the ENTER key; the result appears on the right side of the screen. If you type successive entries such as 7*9 and 6*9, they are all shown. You can then noticed an interesting pattern: the sum of the answer digits is always 9 (i.e., $7 + 2 = 9$, $6 + 3 = 9$, and $5 + 4 = 9$). On simple calculators only one line shows at a time, but on a graphing calculator, results flow down the screen as you work.

```
8*9
          72
7*9
          63
6*9
          54
■
```

Type: 3 ÷ 2 ENTER. Be aware that the symbol on the divide key, ÷, is different from the divide symbol (/) that appears on the screen.

Next, type: 3 − 2 ENTER. Take special care to use the subtraction key on the right side of the keypad. One of the most common errors is interchanging the use of the subtraction key $\boxed{-}$ and the negation key, $\boxed{(-)}$, on the bottom row.

Finally, type: 3 $\boxed{(-)}$ 2 ENTER. Here the negation key creates an error. Give it a moment's thought: subtraction requires two numbers, while negation works on a single number.

```
3/2
          1.5
3-2
          1
3-2
```

Looking carefully at the previous screen shows the two similar symbols: the subtraction symbol is longer and centered, the negation symbol is shorter and raised. This last entry gives you a syntax error as shown here. Error screens replace the work screen and tell you briefly what is wrong. You are forced to respond with 1:Quit or 2:Goto. The Goto option is often the best choice because it will "go to" the error location and allow you to correct it.

```
ERR:SYNTAX
1:Quit
2:Goto
```

> *Tip:* Sequences of calculator screens in an example start and stop with a heavy top and bottom border (as shown above). These borders are meant to mark the beginning and the end of a sequence; this is especially helpful when an example extends over more than one page.

Scientific keys

Next we test the scientific keys. These give values for many expressions used in science, such as the common log (LOG) and natural log (LN) function. The TAN, COS, and SIN keys are the standard trigonometric functions. These functions appear on the screen in lower case followed by a left parenthesis that is automatically inserted for you. First, however, we examine the taking of powers.

We know $2^3 = 8$, and to verify this on our calculator, type

2 ^ 3 ENTER

The often-used square power has the special key x^2. (The x^2 shows only ^{2} on the screen.) So pressing

5 x² ENTER

gives 25 as an answer, the same as 5^2 yields 25.

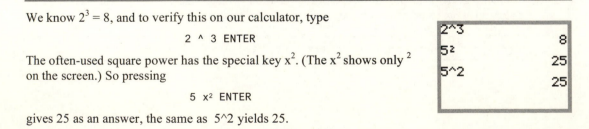

```
2^3
          8
5²
          25
5^2
          25
```

The effect of the x^{-1} key is to take the reciprocal of a quantity, $1/x$. Thus pressing

```
2-1
              .5
log(100)
               2
cos(0)
               1
```

2 x⁻¹ ENTER

gives 0.5 as an answer. To find the value of log 100, type

LOG 1 0 0) ENTER

Trigonometric values can also be found, as shown.

Tip: The right parenthesis in an expression like log(100) is not actually required, but it is a bad habit to leave it off.

Magic tricks to change the keypad

How can we do more with the basic keys? The trick is multiple-state keys. The indicator of the keyboard state is shown by the cursor. The standard is a solid black square. The first change state key is the 2nd key in the upper left corner. After pressing 2nd once, the inside of the cursor shows an up arrow — Presto! — All the keys now have a new meaning. These meanings are indicated just above and to the left of each key, written in the color of the 2nd key. (The color is blue on a TI-84 but yellow on a TI-83.) We have already used the 2nd key to adjust the screen contrast. When the 2nd key is pressed it must be released before pressing the next key. This is not like the shift key on a keyboard, where two keys must be pressed simultaneously.

A note about notation

Some authors use boxed text to specify calculator keys, for example, 2nd . This notation makes reading quite jarring. The TI font alone is sufficient to denote the keys, so boxing is used sparingly in this book. When the desired keystrokes are 2nd followed by OFF, the combination is written together as 2ndOFF. This notation alerts you that you should look for OFF written above the key to be pressed; in this case it is above the ON.

Practicing the 2nd key on the greatest equation ever written

Five symbols, 0, 1, e, π, and i, are frequently used in mathematics. Incredible as it might seem, they are related by a single equation:

$$e^{\pi \cdot i} + 1 = 0.$$

You can practice using the 2nd key for e, π, and i by entering

2ndeˣ 2ndπ * 2ndⁱ) + 1 ENTER

Using ALPHA to store values

The other key that changes the state of keypad is the green ALPHA key; as its name indicates, it is used to enter alphabetic letters such as variables, and it has a secondary role as the Solver activate key. (This feature is discussed in Chapter 2.)

The store key, STO>, appears on the screen as: →. It is used to store a numeric value in a letter variable. To repeatedly use the value of (log(100)+1)/2 in calculations, then you enter these keystrokes

(Remember that pressing the LOG key puts log(on the screen.):

(LOG 100) + 1) / 2 STO> ALPHA_A ENTER

The variable A has a value of 1.5 and can be used in computations, as shown.

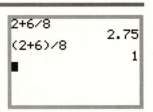

> *Tip:*   The cursor box changes if the ALPHA or 2nd keys are in effect. The ALPHA and 2nd keys work as toggles: if you press either one by mistake and turn on a state, just press the key again to turn it off.

Editing and Recall

We all make mistakes; correcting them on a graphing calculator is relatively easy. Use the arrow keys to navigate the screen and type over your errors.

Corrections in longer expressions

The single most popular error (can errors be popular?) among new users is the failure to use parentheses when needed. In the first calculation of the previous screen, it is vital to get the parentheses in the correct places, if not the answer is different from the one shown. In general, this is a serious problem because the calculator does not stop and alert you with an error screen; instead, it gives you the correct answer to a question you are not asking. There is a prescribed order of operations on your calculator; you can look this up in your *TI–Guidebook* for details if you have questions about this order.

Suppose you want to add 2 and 6 and then divide by 8. We don't need a calculator to tell us the expression has value 1. But if you enter it as $2 + 6 / 8$, the answer is 2.75. You can figure out that the calculator divided 6 by 8 first and then added that to 2. This was not what we wanted. We need to use parentheses to ensure that we are evaluating the correct expression. Try $(2 + 6) / 8$ to get an answer of 1 as expected.

```
2+6/8
              2.75
(2+6)/8
                 1
■
```

> *Tip:*   When you get an unexpected result, go back and check parentheses. Be generous; adding extra parentheses doesn't hurt.

Correction keys: DEL, CLEAR, and 2ndINS

While on the home screen, you can use the arrow keys to move forward and backward. (The up arrow takes you to the beginning of the line, the down arrow to the end.) Press the DEL key to delete the character in the cursor box. Use CLEAR to delete the whole entry line; if the entry line is already clear, then the whole screen clears. Thus, pressing CLEAR twice always clears the screen. If you need to insert one or more characters, you can move (using the arrow keys) to the location at which you want to insert and press the 2ndINS key. The cursor appears as an underline when the insert mode is in effect. Pressing any arrow key turns off the insert mode.

> *Tip:*   Unfortunately, there is no backspace key such as is found on computer keyboards.

Recalling a previous entry

Often we see an error after we have pressed ENTER and left the entry line. The 2ndENTRY key (above ENTER) places the previous entry line on the screen so that it can be edited for the next calculation. In the last screen example, we had the expression 2+6/8 but then realized we needed to add parentheses. After seeing our mistake on the first line, we could have used 2ndENTRY to paste a new copy of the expression on the screen. Then, by using the navigation arrows, we could have inserted the needed parentheses. Note that we would need to use 2ndINS since we were not writing over what was there but inserting new parentheses characters.

Deep recall

By repeatedly pressing the 2ndENTRY key, many of the previous entry lines are accessible. This is called deep recall and is limited to about 128 characters. Using a previous entry overwrites the current entry; you do not go to a new line. A previous entry cannot be inserted into the command line without erasing what was already there — it is all or nothing.

Recycling a previous answer

There is another keystroke-saving feature that is quite handy: the 2ndANS key. The 2ndANS places the variable Ans on the screen and uses the previous answer as its value when calculated.

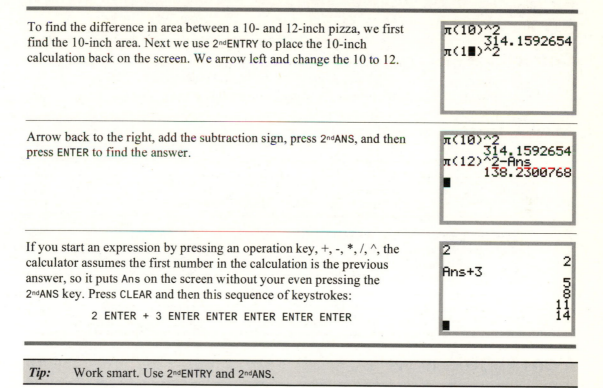

To find the difference in area between a 10- and 12-inch pizza, we first find the 10-inch area. Next we use 2ndENTRY to place the 10-inch calculation back on the screen. We arrow left and change the 10 to 12.

Arrow back to the right, add the subtraction sign, press 2ndANS, and then press ENTER to find the answer.

If you start an expression by pressing an operation key, +, -, *, /, ^, the calculator assumes the first number in the calculation is the previous answer, so it puts Ans on the screen without your even pressing the 2ndANS key. Press CLEAR and then this sequence of keystrokes:

    2 ENTER + 3 ENTER ENTER ENTER ENTER ENTER

Tip: Work smart. Use 2ndENTRY and 2ndANS.

Menus

We have already seen one example of a menu when we committed the error of using the negation sign in place of the subtraction symbol. There were only two choices, but the principle is the same in other cases. Menus are lists of choices that are far too numerous to be available on the keypad.

Press CLEAR twice to start with a fresh screen. Press MATH and you see a menu screen that takes the place of the home screen. Notice that the options are numbered and you can choose by number or by arrowing to a choice and pressing ENTER. All the selections may mean little to you at this point at this point.

```
MATH NUM CPX PRB
1█▶Frac
2:▶Dec
3:3
4:3√(
5:×√
6:fMin(
7↓fMax(
```

The top row shows four menus: MATH NUM CPX PRB. Press the right-arrow key to get the NUM menu. The abs(is our choice, so we press ENTER (or 1).

```
MATH NUM CPX PRB
1█abs(
2:round(
3:iPart(
4:fPart(
5:int(
6:min(
7↓max(
```

Now the home screen reappears and abs(has been pasted onto the home screen. If the home screen is not cleared, the pasting takes place at the location of the cursor, even in the middle of any expression.

```
abs(
```

In case you hadn't guessed, abs() is the absolute value function, which makes numbers positive. Here we see it work. Note that we used the negation sign, not the subtraction key. (Using the subtraction key causes an error.)

```
abs(-6)
            6
```

Tip: Most accomplished users use the number choice as the fastest method of choice in menus. However, if the choice is the first item, then ENTER is best.

Tip: A menu may not show all its items. An arrow in place of the colon indicates that unseen items are available in the arrow's direction.

Tip: Menus are wrap-around so you can press the up arrow once to get to the last item in a list. A choice of menus on the top line wraps from left or right by using the left or right arrows.

A CATALOG of items

An alphabetical list of features is available. The 2ndCATALOG key shows an alphabetic list of items. The list is long, but the black boxed A, in the upper right corner is a reminder that you are in ALPHA mode and you can conveniently jump closer to the desired item by pressing the key of items first letter.

```
CATALOG        A
▶abs(
 and
 angle(
 ANOVA(
 Ans
 Archive
 Asm(
```

Changing the format: MODE

You can control the format of the output for your numerical calculations For example, if you are doing a business application, you might want money answers to come out rounded to two decimal places for the dollar-and-cents format. The MODE screens shown are from the TI-84. The TI-83 screen is the same except for the SET CLOCK feature on the bottom line. This time and date feature has been added to the TI-84.

The MODE key allows you to check and change formats. The default settings are all on the left of the screen so a quick glance tells you if any settings have been changed. Numeric answers are usually shown in FLOAT mode, meaning they show up to eight decimal places if needed.

To change the setting to show dollar-and-cents, use the down arrow to reach the FLOAT line and then use the right arrow to move across to the desired setting. You must now press ENTER to make the change. Press 2ndQUIT to return to your home screen.

We calculate the total cost of a $2.10 Cafe Latte in Seattle where the sales tax is 8.6%. Since we have changed the format to two decimals, our answer is rounded to 2.28 Rounding the result in a display format does not change the stored accuracy. To see this, multiply the answer by 100 and you see that the fuller decimal accuracy is preserved but not displayed. In MODE, reset to FLOAT, before continuing.

For calculus, the RADIAN angle setting is used. This is the default on the MODE menu. For situations where angles in degrees are specified, use the 2ndANGLE menu to paste the degree symbol into the calculation. This is preferable to changing the MODE setting to DEGREE.

Trigonometric calculations are normally in radian measure. Use the 2ndANGLE menu to specify degree or radian measure if needed. Check the entry of angles using a sine function.

The MODE setting choices are described in the *TI–Guidebook*. We mention other settings as we need them, but unless noted otherwise, all our examples assume that the default settings are in effect.

Tip: If your output values are in an unexpected or undesirable format, check the MODE settings. If you are having trouble changing the MODE settings, you may have forgotten to press ENTER before 2ndQUIT.

Notes:

CHAPTER ONE

LINEAR FUNCTIONS AND CHANGE

1.1 Functions and Function Notation

Definition: A *function* is a rule that takes certain values as inputs and assigns to each input value exactly one output value. The use of functions and functional notation is the primary focus of precalculus since it is vital to success in calculus.

In this chapter, we use the calculator to define and evaluate functions, to make tables of values, and to graph functions. In short, we see how to view a function in words, by table, by graph, and by formula (the Rule of Four).

In the previous chapter we used the lower keypad to do our simple numerical calculations. Now we learn to use the Graph/Table keys on the top row, just under the screen. We start with a simple example of a function and show how it is represented as a formula, graph, and table with the aid of the calculator.

> *Tip:* Most examples from this point on are taken from *Functions Modeling Change* by Connolly et al. Page references to these examples are given in the left margin, enclosed by square brackets. For example, see the [2] under the heading below.

Cricket Chirping Rate

[2]   It is a surprising biological fact that most crickets chirp at a rate that increases as the temperature increases. For the snowy tree cricket, the relationship between temperature and the chirp rate is so reliable that this type of cricket is called the thermometer cricket. We can estimate the temperature (in degrees Fahrenheit) by counting the number of chirps in 15 seconds and adding 40. For instance, if we count 20 chirps in 15 seconds, then a good estimate of the temperature is to sum and find $20 + 40 = 60°F$.

Words

This example is already given in words, but we might note that rates are most often given in terms of a single unit, so we can think of the chirp rate in terms of the number of chirps per minute. In this case, since 15 seconds is 1/4 of a minute, we could count the number of chirps per minute and multiply by 1/4.

Formula

From the above discussion we can write a formula for the temperature in terms of the number of chirps:

$$T = \frac{1}{4}R + 40.$$

In this formula we identify R as the independent variable and T as the dependent variable. To enter the formula in the calculator, we translate the independent variable to X and the dependent variable to Y.

> *Tip:* No matter what symbols are used in a function formula, to be used on the TI they must be transformed to have a Y variable equal to an expression using the variable X.

Graph

The next representation is a graph of the function. Like using a camera this requires making a decision about what part is seen. We set a window of values to give us a good picture. Later we find that the calculator also has some preset window commands that act like an automatic focus.

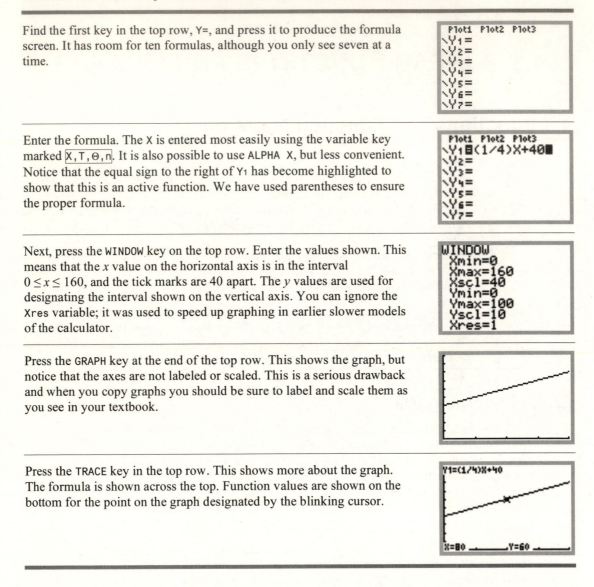

Find the first key in the top row, Y=, and press it to produce the formula screen. It has room for ten formulas, although you only see seven at a time.

Enter the formula. The X is entered most easily using the variable key marked $\boxed{\text{X,T,}\Theta\text{,n}}$. It is also possible to use ALPHA X, but less convenient. Notice that the equal sign to the right of Y1 has become highlighted to show that this is an active function. We have used parentheses to ensure the proper formula.

Next, press the WINDOW key on the top row. Enter the values shown. This means that the x value on the horizontal axis is in the interval $0 \le x \le 160$, and the tick marks are 40 apart. The y values are used for designating the interval shown on the vertical axis. You can ignore the Xres variable; it was used to speed up graphing in earlier slower models of the calculator.

Press the GRAPH key at the end of the top row. This shows the graph, but notice that the axes are not labeled or scaled. This is a serious drawback and when you copy graphs you should be sure to label and scale them as you see in your textbook.

Press the TRACE key in the top row. This shows more about the graph. The formula is shown across the top. Function values are shown on the bottom for the point on the graph designated by the blinking cursor.

Table

The last representation of the function creates a table of values. Like the graph, we set the starting x value and the step size. There are some more advanced features, but they are introduced later.

To make a table of function values the function formula must be first entered as was just done for the graph. Press the 2ndTBLSET key in the top row. Set the initial value of x with TblStart= and the step size with ΔTbl=. For now we ignore the other settings.

To see the table of function values, press the 2ndTABLE key in the top row. The table can be scrolled up or down to see additional values. To see values for $x = 0, 1, 2, 3, \ldots$ reset the ΔTbl=1.

Tip: Recall that to get back to the home screen from a table, press 2ndQUIT.

Tip: Before moving to use advanced features, make sure you know the basic method of entering a formula, graphing a function, and producing a table.

1.2 Rate of Change: Lists and Function Values

The average rate of change tells us about how output values change for a change in input values. For a function $Q = f(t)$ we define

$$\text{Average rate of change } = \frac{\text{Change in } Q}{\text{Change in } t} = \frac{\Delta Q}{\Delta t}$$

Data Lists and Rate of Change

DVD and VCR Sales

[10]   The annual sales (in millions of dollars) of VCR and DVD players between 1998 and 2003 are shown in the table below. What is the rate of change of the annual sales of VCRs and DVDs between 1998 and 2003?

Year	1998	1999	2000	2001	2002	2003
VCR sales	2409	2333	1869	1058	826	407
DVD sales	421	1099	1717	2097	2427	3050

For DVDs, evaluate the quotient, $\dfrac{3050 - 421}{2003 - 1998}$. Enter the quotient on the home screen adding the necessary parentheses:

$$(3050-421)/(2003-1998)$$

To find the average rate of change over many intervals we use a table. Before starting, we note that this is different from earlier examples in which a formula is entered in the Y variable. We use the list feature of the calculator to handle situations where data is given, but no function formula is known. There are two different means of entering a list; both are shown in the following sequence.

Press STAT to see the menu shown. Some screens like this one have related menus shown on the top line (CALC and TESTS), the items in these menus are seen by using the left/right arrows. In Section 1.6 we will use the CALC menu.

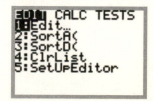

This screen is like a spreadsheet on which to enter numerical lists. There are six numbered lists, L1, L2, ..., L6. Only three at a time show. Data is entered and edited both quickly and easily by using this screen. Remember, just press STAT ENTER. If previous data are in a list, arrow over to use an empty list, or arrow up to the list name and press CLEAR ENTER.

For our VCR and DVD data we enter the year in L1. Put the VCR sales in L2 and the DVD sales in L3. Notice that each entry has a name; the one highlighted is L3(6), meaning the 6th entry in L3.

Use 2ndQUIT to return to the home screen.

With the lists entered it is easy to set up a difference quotient on the home screen to calculate various rates of change. The list names L1 to L6 are easy to enter (and remember) since they are the 2nd key of 1 to 6. Other rates are found by using 2ndENTRY and making editing changes. In this example, after the calculation for VCRs, the DVD values were found by just changing L2 to L3 in two places.

Now that you know about list names and structure, an alternative way of defining a list is on the home screen. Use set brackets { } and commas to create a list and then use the store key, STO>, followed by one of the six list names.

Tip: Do not press ALPHA and L and 4 in hopes of getting the list name L4. The list name is only accessible through the 2ndL4 key, and the number shows as a subscript on the screen.

Rate of Change from Functions

In a similar fashion you can find the rate of change for a function that is defined by a formula. For a function over the interval $a \le t \le b$, we define

$$\text{Average rate of change} = \frac{\text{Change in } Q}{\text{Change in } t} = \frac{\Delta Q}{\Delta t} = \frac{f(b) - f(a)}{b - a}.$$

[14]   We now find the rate of change of the function $f(x) = x^2$ between $x = 1$ and $x = 3$ and between $x = -2$ and $x = 1$.

Function formulas and rate of change

We could easily enter the function formula in the Y= menu, but we use an alternative (and longer) method. On the home screen you can enclose a formula in quotes and use the STO> key to store the formula. (The quote symbol is above the + key.) The hard part is the final pasting of the Y1 symbol. Unlike list names, like L1, there is no keyboard symbol. We must paste Y1 using the following steps.

`"X^2"→█`

Press VARS, then right arrow once to show the Y-VARS menu. From this menu press ENTER (or 1) because we want a function name. The next screen shows the function name list: Y1, Y2,.... Press 1 to choose Y1. See the results on the next screen.

```
VARS Y-VARS
1▊Function…
2:Parametric…
3:Polar…
4:On/Off…
```

The top line shows the definition line after pasting Y1 and pressing ENTER. The word Done confirms that the definition was successful. This longer method was shown because the sequence of pasting a Y variable is used over and over. Now the function is evaluated for $x = 3$, by using the common function notation of putting the input in parentheses. This required pasting our function name.

```
"X^2"→Y1
           Done
Y1(3)
             9
█
```

Once we can evaluate functions, the difference quotient is entered to find the two rates of change. It is most efficient to use 2ndENTRY and be able to use the 2ndINS key to inserting an extra character or DEL to remove one. Notice the extra parentheses to ensure correct calculations.

```
(Y1(3)-Y1(1))/(3
-1)
             4
(Y1(1)-Y1(-2))/(
1-(-2))
            -1
```

Tip: Like list names, you cannot press ALPHA and Y and 1 in hopes of pasting the function name Y1. The function name is only accessible through the sequence of steps: VARS, right-arrow to Y-VARS, ENTER, select.

Tip: Whenever possible, edit long expressions by using 2ndENTRY, 2ndINS, and DEL.

1.3 Linear Functions

Definition: A *linear* function has a constant rate of change. In this chapter we look at what this means in terms of tables, graphs, and formulas.

Linear Depreciation

[19]   The value of new computer equipment is $20,000, and the value drops at a constant rate so that it is worth $0 after five years. If V is the value of the computer equipment t years after the equipment is purchased, find a formula for V in terms of t.

Formula

The basic formula for a linear equation is

$$\text{Output} = \text{Initial value} + \text{Rate of change} \times \text{Input}.$$

From the above discussion we can find the rate of change:

$$\text{Rate of change} = \frac{\Delta V}{\Delta t} = \frac{-20000}{5} = -4000.$$

The initial value is $20,000, so our linear equation is

$$V = 20000 - 4000t.$$

Graph

The next representation is a graph of a linear function. It should be no surprise that the graph looks like a straight line.

First translate variables and write the formula in Y1. Then set the viewing WINDOW to $0 \le x \le 5$ (Xscl=1) and $0 \le y \le 20000$ (Yscl=4000). TRACE shows the equation and graph.

To look at a table, press the 2ndTBLSET and set TblStart=0 and ◻Tbl=1. Press 2ndTABLE to see the table shown. Notice that for each additional year the value drops by 4000.

Tip: The linear equation is written as $y = mx + b$ or as $y = b + mx$, where m is the constant rate of change and b is the initial value.

We next look at a technique to determine if a function is linear by looking only at its table of values. In the previous case we started with a formula; next we start with a data table.

Checking Tables for Linear Functions

[21]   The table below gives values of two functions, p and q. Could either of these two functions be linear?

x	50	55	60	65	70
$p(x)$	0.10	0.11	0.12	0.13	0.14
$q(x)$	0.01	0.03	0.06	0.14	0.15

We want to find out if $\dfrac{\Delta p}{\Delta x}$ and $\dfrac{\Delta q}{\Delta x}$ are constant. We put the data into lists, find the differences within a list, and finally see how to divide the differences to give us a list for $\dfrac{\Delta p}{\Delta x}$ and $\dfrac{\Delta q}{\Delta x}$.

Using ΔList to find differences

We put the three rows of values into the lists L1, L2, L3, by using STAT EDIT ENTER. If these lists already contain values you can either type over the old values or use DEL to delete a single entry. To empty an entire list before entering new data: arrow up to highlight its name, then press CLEAR and ENTER. (Recall 2ndQUIT returns you to the home screen.)

L1	L2	L3	3
50	.1	.01	
55	.11	.03	
60	.12	.06	
65	.13	.14	
70	.14	.15	
------	------	------	

L3(6) =

We want the differences *within* a list to be calculated and stored. We use the home screen command:

ΔList(L1)

The ΔList is a single command pasted from either the CATALOG (under L) or by LIST OPS 7.

```
ΔList(L1)
            {5 5 5 5}
```

Use 2ndENTRY to save a trip back to the CATALOG and change the list name to L2. But this time the list extends off the screen. If you want to see it all, you can scroll back and forth with the right/left arrows.

```
ΔList(L1)
            {5 5 5 5}
ΔList(L2)
{.01 .01 .01 .0…
```

To answer our question about $\frac{\Delta p}{\Delta x}$ and $\frac{\Delta q}{\Delta x}$ we enter the commands shown to determine that p is linear but q is not. Note that this division is done term-by-term, which is important since all the Δx values from a table may not always be the same as they are in this example.

```
ΔList(L2)/ΔList(
L1)
{.002 .002 .002…
ΔList(L3)/ΔList(
L1)
{.004 .006 .016…
```

Tip: Be careful about assuming too much from a few table values.

Warning about Linear Looking Graphs

A graph may look linear, but you should not automatically assume that it is linear. Fooled by the local appearance of flatness, for years nearly everyone assumed that the world was flat. In the following example a function is defined whose graph looks linear in a small time frame, but the function is not linear.

[23]   The function $P = 100(1.02)^t$ models the population of Mexico in the early 2000s. Here P is in millions, and t is the number of years since 2000.

First translate variables and write the formula in Y1. Then set the viewing WINDOW to $0 \leq x \leq 5$ (Xscl=1) and $0 \leq y \leq 150$ (Yscl=50). TRACE shows the equation and graph.

Reset the viewing WINDOW to $0 \le x \le 60$ (Xscl=10) and $0 \le y \le 300$ (Yscl=100). Use TRACE to show the equation and graph. Now we see that over a longer time period the graph is not linear.

Tip: Table values are more reliable here. They would show P is not linear over $0 \le x \le 6$.

1.4 Formulas and Graphs for Linear Functions

In this section we look at some features of the graphing calculator that aid in getting a good viewing window for a graph.

Soda and Chips Example

[30]   We have $24 to spend on soda and chips for a party. A six-pack of soda costs $3 and a bag of chips costs $2. The number of six-packs we can afford, y, is a function of x, the number of bags of chips we decide to buy. Find an equation relating x and y; graph the equation and interpret the intercepts and other x, y values.

If we buy x bags of chips we have spent $2x$ dollars, if we buy y six-packs of soda we have spent $3y$ dollars, and the total is $24. An equation relating x and y is:

$$2x + 3y = 24.$$

However, this form of a linear equation cannot be entered into the calculator. We solve for y and find

$$y = 8 - \frac{2}{3}x.$$

In all the previous examples, the viewing window has simply been given. The decision on a viewing window is the most difficult part of graphing.

Finding a Good Viewing Window

After entering Y1=8-(2/3)X, press the ZOOM key to show the ZOOM menu shown. Select option 6:ZStandard to set the viewing window to $-10 \le x \le 10$ (Xscl=1) and $-10 \le y \le 10$ (Yscl=1). This is a good "first take" window for functions involving small numbers.

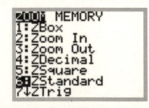

We see that the y-intercept is 8. This makes sense because if we buy 8 six-packs, we have no money left for chips. If we buy all chips, we know we can buy 12 bags. (This point is not in the window.) Also, in this situation, negative values are meaningless. So our viewing window can be improved.

After the above reasoning, we set the viewing window values as shown. Note that we have set the Ymax value to be greater than the *y*-intercept so that the formula at the top of the screen does not obscure the intercept. (You could also set Ymin=–2 to leave room for values at the bottom.)

Now TRACE gives a good viewing window. Viewing windows are a matter of taste and may also depend on what you want to do with them. For example, if you want to use TRACE and see how many six-packs to buy for two or three bags, then this window is lousy. (Try tracing to exactly *x* = 2 or *x* = 3.)

There are ZOOM options that reset screens in various ways. The 8:ZInteger resets the window so that the X values are integers when traced. (Just what we need in this case.) Beware: this is a two-step process. After selecting 8:ZInteger, the cross-hair cursor blinks on the screen, waiting for you to select a new window center. It hardly matters in this case, so just press ENTER. Now press TRACE, and integer values of *x* are shown when tracing. Notice, however, that our viewing window is not as good as the previous one.

Can we have our cake and eat it too? Return to the previous settings with ZOOM MEMORY 1:ZPrevious (or re-enter them manually.) Press TRACE 2. (This shows *x* and *y* with *x* = 2.) After these steps the screen should appear as shown.

Press ENTER and you see that trace has gone directly to the exact value you wanted.

Tip: For a single value of the function, the above method is quite handy, but if you want to see many values, say from 0 to 12, then a table is the best choice.

ZOOM Options

1:ZBox   Resets the window after the box is determined. This allows you to draw a rectangular box on the screen, which becomes the new viewing window.

2:Zoom In   Resets after centering. Move the cursor to the center of where you want to look closer and press ENTER to see a close-up view.

                  (The scale of zooming in and out is set in ZOOM MEMORY 4:SetFactors...).

3:Zoom Out   Resets after centering. Move the cursor to the center of where you want to back away from and press ENTER.

`4:ZDecimal`	Sets the viewing window to the seemingly bizarre values $-4.7 \leq x \leq 4.7$ (Xscl=1) and $-3.1 \leq y \leq 3.1$ (Yscl=1). To view a function close to the origin, this setting has the advantage of tracing x values that are the exact decimal values of -4.7, -4.6,…, 0, 0.1, …, 4.6, 4.7. In other words, tracing would evaluate 1.5 exactly, not some long decimal close to it.
`5:ZSquare`	Resets the aspect ratio of the window. Since the screen is not square some window settings make circles look like ellipses or perpendicular lines may not appear to have right angles. *Note*: ZDecimal is already a square setting.
`6:ZStandard`	Sets the viewing window to the values $-10 \leq x \leq 10$ (Xscl=1) and $-10 \leq y \leq 10$ (Yscl=1).
`7:ZTrig`	Sets the window to nearly $-2\pi \leq x \leq 2\pi$ but it has an exact Xscl=$\pi/2$. This is useful for tracing trigonometric function values. We will use this in Chapter 6.
`8:ZInteger`	Resets, after centering, to trace integer values.
`9:ZoomStat`	Resets after calculating a suitable window for a data list graph. See Section 1.6.
`0:ZoomFit`	Resets using the current X values and finds suitable Ymin and Ymax values.

Tip: Notice that each description of a zoom option starts with set or reset in the above list. Set gives the same window every time, but reset makes an adjustment based on the current window. You can expect a cross-hair cursor to appear for boxing and centering. Normally, center readjustment is unnecessary, and pressing ENTER moves you along quickly.

Viewing Windows for Applications

In closing this section, we note that the set zoom keys (4, 6, and 7) are often good for the kind of formulas that need to be graphed in abstract mathematics, such as $y = 2x + 3$ or $y = x^2$. But in actual applications the input and output values are often large or restricted (or both), and the zoom window is inappropriate. In most applications the input values, the X settings, are known.

[23]   For example, in Section 1.3 we saw the population of Mexico function, $P = 100(1.02)^t$, and we knew the input values. We wanted to look in the short-run, 0 to 5 years, and then in the long-run, 0 to 60 years. It was less clear what y values should be used for the viewing window. One simple way to find them is first to do a table and see what kinds of output values need to be accommodated. A quicker way to get y value window settings is with the `0:ZoomFit`.

First enter the function, or select it, in the Y= menu. Press WINDOW and set just the X values as shown. It makes no difference what the Y values are.

```
WINDOW
 Xmin=0
 Xmax=5
 Xscl=1
 Ymin=0
 Ymax=10
 Yscl=1
 Xres=1
```

We can press ZOOM 0, if we remember, or we can arrow down until we highlight that selection. (*Helpful hint*: The menus are wraparound, so that by pressing ZOOM and up-arrow, you are immediately at `0:ZoomFit` because it is the last option.)

```
ZOOM MEMORY
4↑ZDecimal
5:ZSquare
6:ZStandard
7:ZTrig
8:ZInteger
9:ZoomStat
0↑ZoomFit
```

We do get a perfect fit of y values — we see both the smallest and the largest — but it is not a very good window. However, by pressing WINDOW, we can see what those y values are. Note that the lack of a solid bottom line indicates that the x-axis is not shown.

Now graph the population function again for $0 \le x \le 60$ using the ZoomFit feature. As in Section 1.3, this graph tells us that the function is not linear.

1.5 Geometric Properties and Multiple Linear Functions

In this section we see how to graph multiple functions. The main emphasis is on finding the intersection point of two lines.

The Effect of Parameters

In the formula $y = b + mx$, we recognize x as the input and y as the output, but what are b and m? These are some constant values that are set before we work with the equation. We call these *parameters*. In the following example, we see that b tells us where the line crosses the y-axis and m tells us about the inclination of the line.

In the Y= screen we define three functions Y1=1+X, Y2=2+X, and Y3=3+X.

Press ZOOM ZStandard to see the three functions graphed in the same window.

Using graph styles

Multiple function graphs may need a graph style to make them more distinguishable. By arrowing to the far left of a Y= equation definition and pressing ENTER seven times you can cycle through graph styles. Style 3 and 4 are shading above and below. Style 5 and 6 uses a bubble cursor when graphing the curve.

See the *TI-Guidebook* for more details.

Define the three functions as shown. By arrowing to the far left of Y2 and pressing ENTER you can change to the thick graph style for Y2.

The thick style assigned to Y2 to make it easily distinguishable from the other two lines.

Tip: Using the dotted style does not always show differently from the standard line style.

Tip: Clearing a function definition automatically resets the graph style to the standard line.

Using a List in a definition: a parameter trick

This is a cheap magic trick, but it's fun. In the Y= screen we define the function Y1={1,2,3}+X and use a ZStandard window and TRACE. Hopefully, you see through the trick. This single formula defines three functions: $y = 1+x$, $y = 2+x$, and $y = 3+x$.

Turn off (deselect) Y1 and define the function Y2={1,2,3}X. Here again use a ZStandard window. You should be able to explain this screen based on the previous screen.

Tip: There is a disadvantage in using the above magic trick. If you use TRACE, the cursor still jumps from graph to graph, but the label does not change so you may not know which function formula you are actually on.

Horizontal and Vertical Lines

[38]   It is simple to show a horizontal line: just set the Y variable to some number. However, a vertical line is not a function and thus cannot be defined via the Y= menu. You can draw vertical lines on an existing graph as shown below.

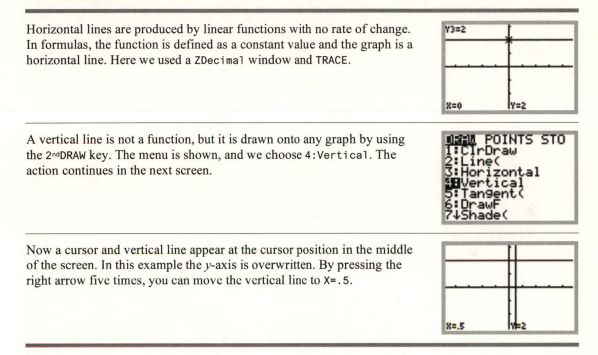

Horizontal lines are produced by linear functions with no rate of change. In formulas, the function is defined as a constant value and the graph is a horizontal line. Here we used a ZDecimal window and TRACE.

A vertical line is not a function, but it is drawn onto any graph by using the 2ndDRAW key. The menu is shown, and we choose 4:Vertical. The action continues in the next screen.

Now a cursor and vertical line appear at the cursor position in the middle of the screen. In this example the *y*-axis is overwritten. By pressing the right arrow five times, you can move the vertical line to X=.5.

Parallel and Perpendicular

Parallel lines have the same slope (rate of change). All three functions $y = 1+x$, $y = 2+x$, and $y = 3+x$ have a slope of 1, and so they are parallel in the first graphing sequence of this section. Perpendicular lines have slopes that are the negative reciprocals of each other, but depending upon the viewing window the TI may give you a distorted view.

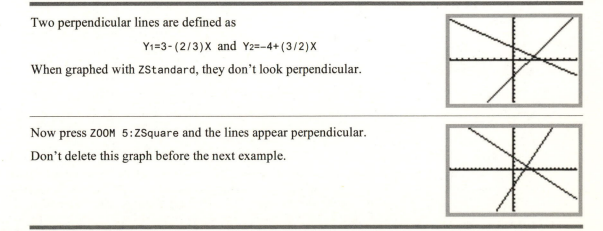

Two perpendicular lines are defined as

$$Y_1=3-(2/3)X \text{ and } Y_2=-4+(3/2)X$$

When graphed with ZStandard, they don't look perpendicular.

Now press ZOOM 5:ZSquare and the lines appear perpendicular.

Don't delete this graph before the next example.

Finding Points of Intersection

It is a common need to find the intersection of graphs. We start with finding the intersection point of the two lines in the last example and then go on to find multiple intersections for an application.

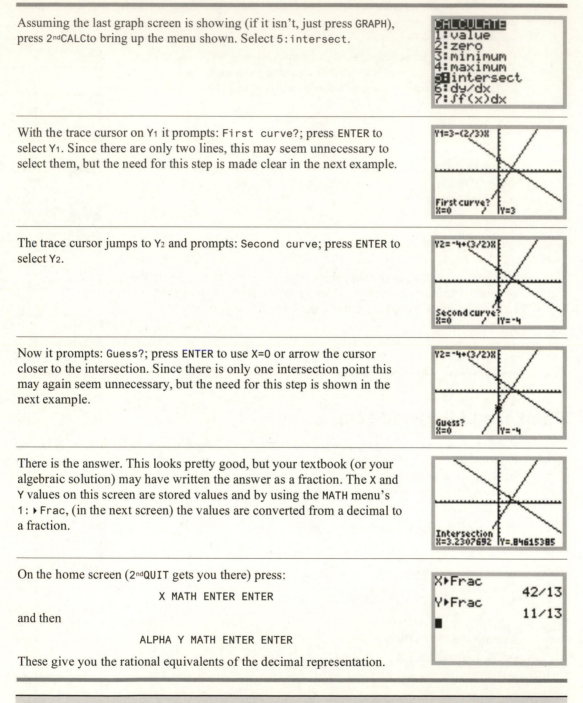

Assuming the last graph screen is showing (if it isn't, just press GRAPH), press 2ndCALCto bring up the menu shown. Select 5:intersect.

With the trace cursor on Y₁ it prompts: First curve?; press ENTER to select Y₁. Since there are only two lines, this may seem unnecessary to select them, but the need for this step is made clear in the next example.

The trace cursor jumps to Y₂ and prompts: Second curve; press ENTER to select Y₂.

Now it prompts: Guess?; press ENTER to use X=0 or arrow the cursor closer to the intersection. Since there is only one intersection point this may again seem unnecessary, but the need for this step is shown in the next example.

There is the answer. This looks pretty good, but your textbook (or your algebraic solution) may have written the answer as a fraction. The X and Y values on this screen are stored values and by using the MATH menu's 1:▸Frac, (in the next screen) the values are converted from a decimal to a fraction.

On the home screen (2ndQUIT gets you there) press:

X MATH ENTER ENTER

and then

ALPHA Y MATH ENTER ENTER

These give you the rational equivalents of the decimal representation.

Tip: When using ▸Frac, if a decimal value cannot be converted, or the resulting denominator has more than three digits, then the decimal value is shown.

Which car agency is cheaper?

[37]   The cost, C, in dollars of renting a car for a day from three different rental agencies and driving the car d miles is given by the following three functions:

$$C_1 = 50 + 0.10d \qquad C_2 = 30 + 0.20d \qquad C_3 = 0.50d \,.$$

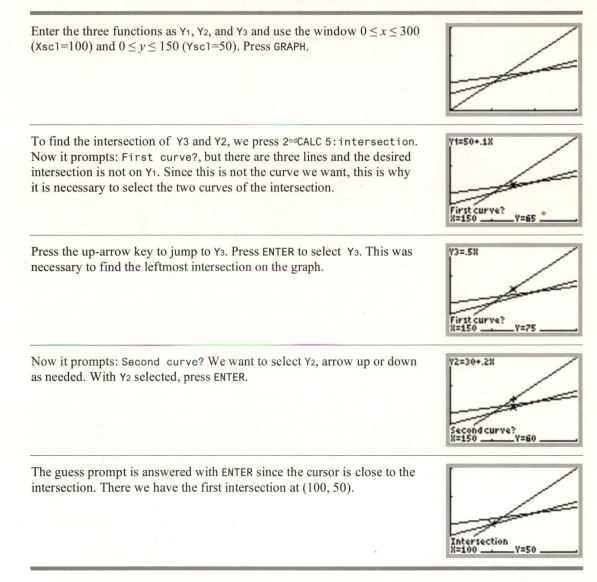

Enter the three functions as Y_1, Y_2, and Y_3 and use the window $0 \le x \le 300$ (Xscl=100) and $0 \le y \le 150$ (Yscl=50). Press GRAPH.

To find the intersection of Y_3 and Y_2, we press 2ndCALC 5:intersection. Now it prompts: First curve?, but there are three lines and the desired intersection is not on Y_1. Since this is not the curve we want, this is why it is necessary to select the two curves of the intersection.

Press the up-arrow key to jump to Y_3. Press ENTER to select Y_3. This was necessary to find the leftmost intersection on the graph.

Now it prompts: Second curve? We want to select Y_2, arrow up or down as needed. With Y_2 selected, press ENTER.

The guess prompt is answered with ENTER since the cursor is close to the intersection. There we have the first intersection at (100, 50).

The other intersections are found at (125, 62.5) and (200, 70). From this we conclude that agency three is cheapest up to 100 miles, agency two is cheapest between 100 and 200 miles, and agency one is cheapest for more than 200 miles. As it turns out, we didn't actually need to find the intersection at 125.

1.6 Fitting Linear Functions to Data

Laboratory Data: Viscosity of Motor Oil

[44]   The viscosity of a liquid (its resistance to flow) depends on the liquid's temperature. Viscosity of motor oil is a measure of its effectiveness as a lubricant in the engine of a car. Thus, the effect of the engine temperature is an important determinant of motor oil performance. See the following table for example data.

T, temperature (°F)	v, viscosity (lbs·sec/in^2)
160	28
170	26
180	24
190	21
200	16
210	13
220	11
230	9

Find a formula that can be used to approximately predict the viscosity of the oil at a given temperature.

Graphing a Scatter Plot

Press STAT ENTER to enter the list editor. If necessary clear previous data in the list. Enter the values from the data table. Note that not all the values show on the screen at one time. This screen and the next require care to distinguish between the keys labeled STAT and STATPLOT, the latter is a 2nd key above the Y= graph key.

Press 2ndSTATPLOT to show a menu of the three special plots called Plot1, Plot2 and Plot3. In this screen we see they are all turned Off. We need to set the first one to On.

We select the first plot by pressing 1, and see a setup screen that allows us to make all the graph settings (except for the window). The settings are changed by arrowing to a choice and pressing ENTER. First turn the plot On. The Type icon shown is a scatter plot. The Xlist and Ylist are set by pasting in L1 and L2. The Mark for each point is a square.

Press Y= to return to the function definition screen. On this screen you see Plot1 highlighted. Turn off any functions that are unrelated to this plot.

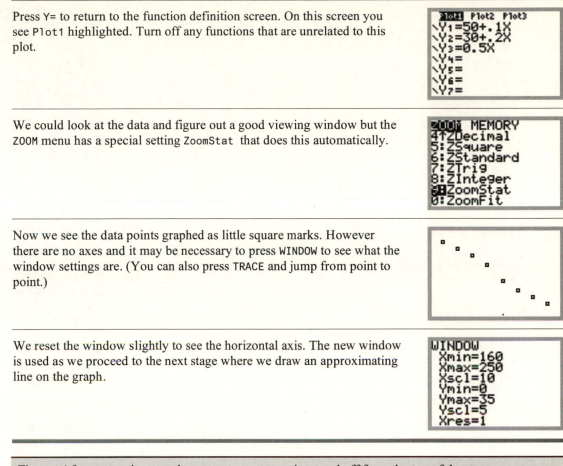

We could look at the data and figure out a good viewing window but the ZOOM menu has a special setting ZoomStat that does this automatically.

Now we see the data points graphed as little square marks. However there are no axes and it may be necessary to press WINDOW to see what the window settings are. (You can also press TRACE and jump from point to point.)

We reset the window slightly to see the horizontal axis. The new window is used as we proceed to the next stage where we draw an approximating line on the graph.

Tip: After a Plot is properly setup, you can turn it on and off from the top of the Y= menu.

Regression Curves

A regression curve is a function that approximates the data. In this case we see that the data values approximate those of a linear function. The approximating line is called a regression line.

In the coming section, your screen may not appear exactly like the one shown. To avoid dealing with this in the middle of the example we now make a settings change. Start with a clean home screen and use 2ndCATALOG to paste in DiagnosticsOn Press ENTER, and Done appears. Once this is done, this setting stays on unless it is deliberately reset to be off.

Finding a regression line

This screen is our scatter plot using the window from the previous plot.

Press STAT and arrow to the CALC menu. Make the choice
4:LinReg(ax+b). Choices 4 and after are the different kinds of regression
curves that you can use. We use other regression curves in Section 3.3
and Section 9.7.

After pasting LinReg(ax+b) on a clean home screen, we specify the two
lists and (optionally) a Y to write the regression equation into. We are
using L1 and L2 and pasting the equation into Y1. Press ENTER to see the
screen in the next frame.

This gives the equation of the regression line, but it also has other
information. The r is called the correlation coefficient, and it tells you
the closeness of the approximation equation. See your text for more
information.

We press TRACE to see the scatter plot and the regression line. Press the
up arrow to see that at the top of the screen the regression equation has
been pasted to Y1 with full accuracy.

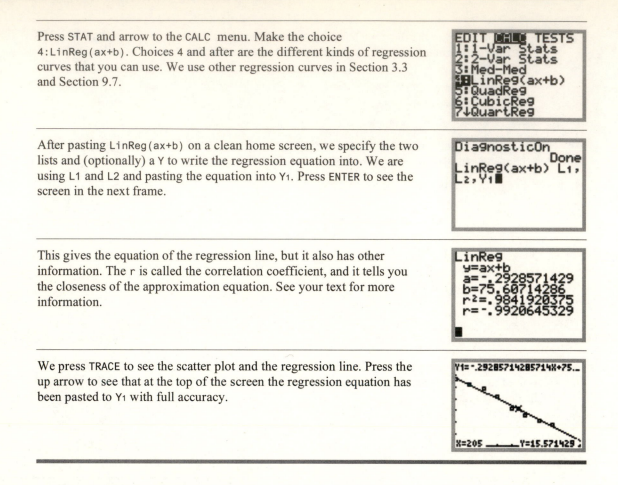

Interpolation and Extrapolation

So what is the use of a regression line? It is a compact formula that can be used to approximate the
data values. When it approximates a value between known values it is said to be an interpolation. If the
value is before or after known values, then it is called an extrapolation.

When tracing the regression line, a value for X can be entered to find a
value for Y. We find an interpolation value for X=196.

Arrow right to find the value near X=240. This is an extrapolation.

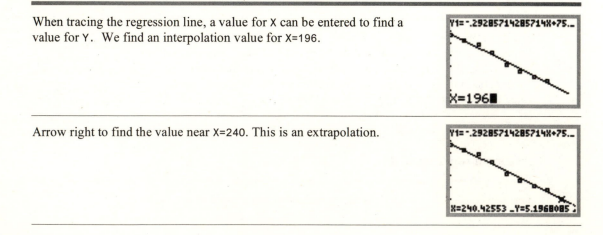

If you decide you would like to see a value for X=300, which is not on the screen, then you get an error.

To remedy this error, expand the window or use just the TABLE feature.

Notes:

CHAPTER TWO

FUNCTIONS

2.1 Input and Output

For a function, each input has a unique output. For functions defined by a formula we now review how to find this output for a given input. This is called *evaluating the function*. Next we reverse the process: for a given output we find an associated input value. In the equation $y = f(x)$, this means that we know y and want to find x. This reverse process is called *solving the equation*.

Evaluating a Function Using a Formula

Recall from Section 1.2 that when a function is defined using the notation $y = f(x) = x^2$, then $f(10) = 10^2$ and we write $f(10) = 100$. In words: "We are evaluating the function at 10." For example, to evaluate the function f using the TI, define Y₁=X^2 using the Y= menu (or from the home screen), and then enter Y₁(10) on the home screen. The mathematical notation $f(10) = 100$ is translated to the TI notation Y₁(10) with answer 100.

Numerical vs. algebraic evaluation

[62]   Let's work on a more challenging function

$$g(x) = \frac{x^2 + 1}{5 + x}.$$

Evaluate the following expressions

(a) $g(3)$ (b) $g(-1)$ (c) $g(a)$ (d) $g(a-2)$.

First translate variables and write the formula in Y₁. Note the necessity of parentheses to force the correct order of operations in the fraction. Use 2ndQUIT to return to the home screen.

```
Plot1 Plot2 Plot3
\Y1◼(X^2+1)/(5+X
)
\Y2=
\Y3=
\Y4=
\Y5=
\Y6=
```

We can evaluate the function Y₁ at 3 and −1. (In Section 1.2 we learned how to paste the Y₁ to the home screen by using the VARS key.) After finding Y₁(3) you can use 2ndENTRY to save a trip back to the VARS menu when finding the second function value.

```
Y1(3)
              1.25
Y1(-1)
                .5
```

To find $g(a)$, we might hope that Y₁(A) would give an algebraic answer in terms of A, but the TI operates only at the numerical level. Thus the expression Y₁(A) gives the value of the function using the current value of A. Look at your current value of A (which may vary from the one shown) and then evaluate the function to verify this.

The more advanced calculators like the TI-89 and Voyage 200 have symbolic capabilities that actually evaluate part (c) and part (d) algebraically. By using algebra or a TI-89, you can confirm:

$$g(a) = \frac{a^2+1}{5+a} \quad \text{and} \quad g(a-2) = \frac{a^2-4a+5}{3+a}.$$

Tip: Evaluating a function from the graph screen is possible by pressing 2ndCALC 1:value and giving an X value between the Xmin and Xmax of the viewing window.

Finding Input Values from Output: Solving Equations

[63]   Suppose $y = \dfrac{1}{\sqrt{x-4}}$.

(a) Find an x-value that results in $y = 2$.

(b) Is there an x-value that results in $y = -2$?

Using Solver

Part (a) is asking us to solve the equation

$$2 = \frac{1}{\sqrt{x-4}}.$$

We could do this algebraically by hand, but the TI does the job as shown below. But first, to get our bearings, we graph the function.

Graph the function in a ZStandard window. Use TRACE and arrow over as shown. (In the next section we talk about the strangeness of this graph.) The cursor point has a value of y close to 2, so that the values of x we are looking for is about 4. Thus, we have approximately solved for x.

We now find a more accurate solution. Press MATH 0 as shown to begin the Solver.

An equation line appears on which to complete the eqn:0= prompt for the equation. Notice that we have to transform the equation by moving all the terms to the right. (If some previous equation is shown, use CLEAR before starting the new equation entry.) Press ENTER or the down arrow when the equation is entered.

The X value from the graph is shown as a guess. Now press ALPHA SOLVE. (SOLVE is above the ENTER key.) This command says: solve the equation with the guess shown. The bound= line is used to restrict the *x* values it looks for, but we ignore that line for now.

Once solved, the solution is shown with a black square block to the left of the freshly calculated variables. Do not trust a Solver screen until you see this symbol. The additional line left-rt=0 indicates that this solution is exact. When this line is not zero it alerts you that the solution is an approximation by giving the difference between the left and right side of the equation.

The difference between Solver and solve

Solver is a new and improved version of the solve command from the older TI-82. If you are using a TI-82 you need to translate all Solver into solve. The solve command of the TI-84/83 is not on any menu but it is in the CATALOG. The syntax for solve is:

$$\text{solve}(expression, variable, guess).$$

The result is a value of the *variable* that makes the *expression* zero by starting to search for that value at *guess*. The previous Solver equation solution is redone below using solve.

From the home screen, use 2ndCATALOG to paste solve(and complete the command as shown. We have used a guess of 5 because we know from the graph that the function is undefined for $x \le 4$. A common blind guess of zero is frequently used in the solve command, but in this case it would cause an error. Such an error is shown in the next subsection.

Tip: Solver and solve allow you to enter expressions containing the Y= variables. For example, solve(2-Y$_1$,X,5) is the same as the above screen's command, when Y$_1$=1/√(X-4).

Solver error warnings

Because they use numeric techniques, solve and Solver are in some cases unreliable. In the first example below they fail to find a solution that actually exists. The second example explains the error message given when you try to find a solution that doesn't exist. This answers our original problem's part (b): Is there an *x*-value that results in $y = -2$?

In Solver, first check to see that the proper equation is entered. (This step not shown, but the proper equation is shown on the top line of the screen.). Now we attempt to re-create the known solution with a guess of X=0. The surprise is that we get an error message.

This error message should not be interpreted as telling us that there is no real answer. We already know there is one! By starting with $x = 0$, the value $\sqrt{(-4)}$ is encountered which is nonreal.

Now we attempt to answer part (b) by using $y = -2$. In this case the function is always positive and thus could never be equal to -2. There is no x value that can be found, but the error message doesn't explicitly say that, as shown on the next screen.

The `Solver` method in finding a zero is to find one positive and one negative value of the difference, then it squeezes in on the zero that is trapped between. In cases where all the values of the function are positive it eventually quits and says: NO SIGN CHNG, (no sign change.)

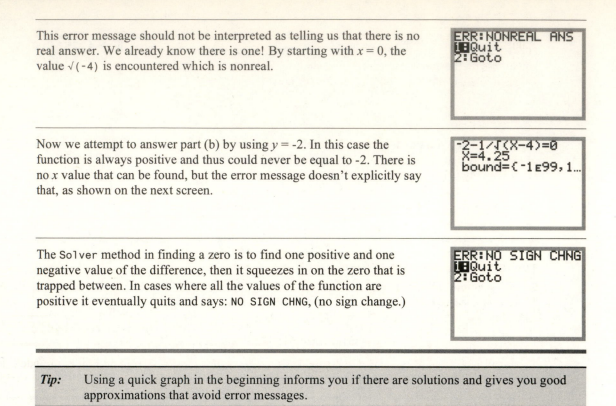

Tip: Using a quick graph in the beginning informs you if there are solutions and gives you good approximations that avoid error messages.

2.2 Domain and Range

A good viewing window for a function's graph provides broad hints about the domain and range of the function. This section focuses on finding good viewing windows and interpreting the results about the domain and range.

Sunflower Growth

[70]   A sunflower plant is measured every day t, for $t \geq 0$. The height, $h(t)$ centimeters, of the plant is modeled by using the function

$$h(t) = \frac{260}{1 + 24(0.9)^t}.$$

What is the domain of this function? What is the range?

Enter Y₁=260/(1+24*(.9)^X), check your function definition for correct parentheses. A common-sense estimate of the growing season is 90 days. We are given the restriction $t \geq 0$; thus we can graph for $0 \leq t \leq 90$. In WINDOW, set Xmin=0 and Xmax=90. Use ZoomFit to give a good first window.

The axis values are not shown, so press WINDOW to see the range of values. Our answer would be that for a domain of $0 \leq t \leq 90$ the range is $10.4 \leq h(t) \leq 259.53$. The problem said that the model could be used for $t \geq 0$. Although it may not make sense in this model, let's see what happens to the $h(t)$ values, as t gets larger and larger.

Use TRACE to look at values of Y₁ for X larger than 90. Hold down the right-arrow key and the cursor moves to the right until it reaches 90. At that point it *pans* (shifts) over and show values greater than 90. In the screen shown, the cursor has disappeared, but the values are read at the bottom of the screen.

The WINDOW settings $0 \leq x \leq 90$ and $-50 \leq y \leq 300$ produces the screen shown. (Extra room is given so that tracing information does not obscure the top or bottom of the screen.) Now by panning to the right, you can convince yourself that the Y values get closer and closer to 260.

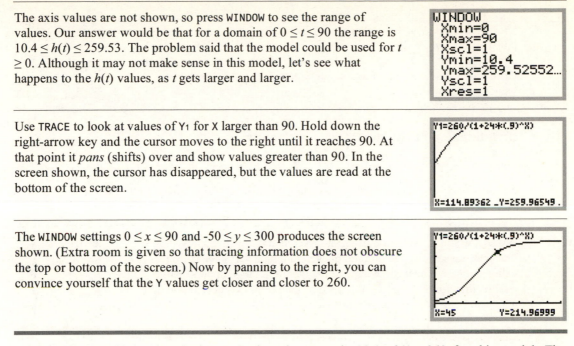

In summary, if the domain is $t \geq 0$, then the range is $10.4 \leq h(t) < 260$ for this model. The mathematical domain of this function is not restricted by reality, it is all real numbers. This is verified when we discuss exponential function in the next chapter.

Tip: In general, when setting a WINDOW make the Y settings generous to accommodate for the top and bottom information in TRACE.

Undefined Points

[71]   A single undefined point in the domain of a function is often hard to see on a graph; however, undefined intervals are easier to spot. We return to our strange graph from Section 2.1,

$$y = \frac{1}{\sqrt{x-4}}.$$

Because we can only take the square root of a positive number, we see that this function is undefined for $x < 4$. In addition, if $x = 4$, then we have a zero in the denominator which is also not defined. Thus our domain is $x > 4$. We now see how the TI communicates these conclusions.

Undefined domain values

Using ZStandard and TRACE, we see our graph. A very strong hint that the function is undefined is the fact that at X=0 there is no Y value given. Repeated pressing of the right-arrow key shows that Y remains undefined until you reach points with x greater than 4.

We check this out by using a table. In a table, an undefined domain value has ERROR as the Y values. (See $x = 4$.) The table increment was set as 0.001 so that we could see the numeric behavior close to 4. Reading the table values upward toward X=4, the values of Y1 are getting larger and larger. You might repeat this table with increment 0.0001 to see the Y1 values get even larger.

X	Y1	
4	ERROR	
4.001	31.623	
4.002	22.361	
4.003	18.257	
4.004	15.811	
4.005	14.142	
4.006	12.91	

X=4

Undefined points on the domain and range

[71] Suppose we have the function $y = \dfrac{1}{x}$. We see right away that the denominator is x, so that it is not defined for $x = 0$. Algebraically, we can also show that the range is all real values except $y = 0$. The graph gives us hints that we are correct.

We first graph the function using ZStandard and TRACE; then we interpret the screen. The domain does not include X=0 since there is no Y= value given. We next use a table to decide if the range includes large positive/negative values since it disappears into the vertical axis. There may also be a break in range values at $y = 0$.

Y1=1/X

X=0   Y=

We first investigate the behavior close to zero. We see that table values seem to be getting larger as they get closer to zero. We can use even smaller ΔTbl values to confirm that the y values are unbounded.

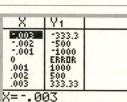

X	Y1	
-.003	-333.3	
-.002	-500	
-.001	-1000	
0	ERROR	
.001	1000	
.002	500	
.003	333.33	

X=-.003

Algebraically we know there is no x such that $1/x = 0$. The graph shows that if there were such a value it would be quite large (some x off the screen.) So we use a table with ΔTbl=100 (or some other large value) to see the y values get close but never reach $y = 0$.

X	Y1	
-300	-.0033	
-200	-.005	
-100	-.01	
0	ERROR	
100	.01	
200	.005	
300	.00333	

X=-300

We conclude that this function has a domain of all real numbers except $x = 0$ and a range of all real numbers except $y = 0$. The graph and/or table did not specifically state that this is the case. To be absolutely sure, we must make an algebraic argument. The TI is used to make a good guess at the domain and range. Or, after having found the domain and range algebraically, the calculator serves as a confirmation.

2.3 Piecewise Defined Functions

In real life there are many piecewise defined functions, for example, the rule used to calculate our income tax. This tax function is published as a table that gives the tax due for an income bracket. A piecewise defined function is a collection of two or more rules, and the particular rule used for output, depends on the value of the input. As another example, if you are boiling water a quart of water on your stove, the temperature of the water depends upon the number of minutes since you started heating it. Thus, there is a formula to model the temperature before reaching the boiling point, and a second formula that goes into effect for the time after the boiling point has been reached. (Remember: After the boiling point is reached the temperature is constant.)

The Absolute Value Function

[76]   A piecewise definition for the absolute value often has the form

$$f(x) = \begin{cases} x & \text{for} \quad x \geq 0 \\ -x & \text{for} \quad x < 0. \end{cases}$$

One rule is for x greater than or equal to zero, and another rule is for x less than zero.

Let's graph this function. From the Y= menu, use 2ndCATALOG to paste in the abs function.

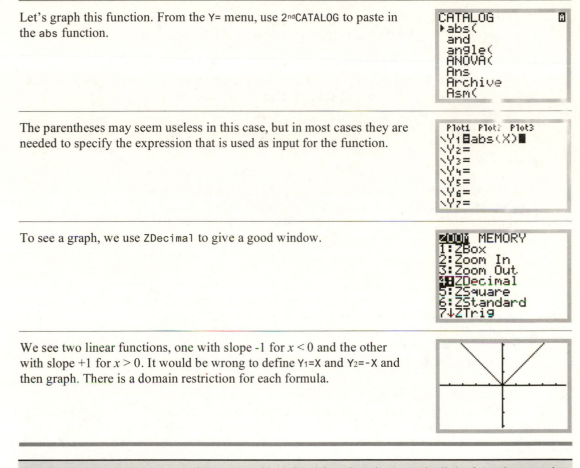

The parentheses may seem useless in this case, but in most cases they are needed to specify the expression that is used as input for the function.

To see a graph, we use ZDecimal to give a good window.

We see two linear functions, one with slope -1 for $x < 0$ and the other with slope +1 for $x > 0$. It would be wrong to define Y1=X and Y2=-X and then graph. There is a domain restriction for each formula.

Tip: A common error in graphing piecewise defined functions is to graph all the formulas over the complete domain.

Logical Expressions

The domain is usually designated by intervals, such as $x > 0$. If $x > 0$ is wrapped in parentheses it is called a logical expression and the calculator can check to see if it is True or False. A True expression has a value of 1; a False expression has a value of 0. To define a piecewise function, we introduce a little trick that use logical expressions to restrict each formula by including its value only on its particular interval. For example, the absolute value function could be written as

$$f(x) = -x\,(x < 0) + x\,(x \geq 0).$$

This works because when the logical is False then that part of the sum is zero, but when it is True then the logical is 1 and we have the formula's value.

[73]   Let's do a more difficult example. Graph the function

$$g(x) = \begin{cases} x+1 & \text{for} \quad x \le 2 \\ 1 & \text{for} \quad x > 2 \end{cases}$$

For this example, it is helpful for you to read the first three steps below before starting to enter anything in your calculator.

The definition of g is translated into

$$g(x) = (x+1)(x \le 2) + (1)(x > 2),$$

where the two inequalities act as logical expressions. We start to enter the function in Y₁ but come to a point where we now need the logical symbol $\le$.

Logical symbols are listed in the 2ndTEST menu. Here we paste the $\le$ symbol from the 2ndTEST menu. Continue to enter the whole expression.

Here is how it works: If $x \le 2$, the expression (X≤2) has a value of 1 but (X>2) has a value of 0. This means that Y₁ is Y₁=(X+1)(1)+(1)(0), which is $y = x + 1$. But if $x > 2$, the expression (X≤2) has a value of 0, but (X>2) has a value of 1. Now Y₁=(X+1)(0)+(1)(1), which is $y = 1$.

Graph using ZDecimal and get a little surprise, if you are using a TI-83. It has connected the values on either side of $x = 2$. When the TI graphs a function, it evaluates 95 x-values to plot 95 points. These points are normally connected. In this case the connection on either side of $x = 2$ should not be made. This situation has been corrected on the TI-84.

To change this on a TI-83, go into the MODE menu, and arrow down and over to Dot. Press ENTER to make the selection. Press 2ndQUIT to leave the MODE screen. For most piecewise defined functions, this is the best setting.

Graph again to see the difference. TI-84 users saw this screen on the first graphing.

A Table to Draw Open/Closed Circles on Graphs

Note that if you copy the previous graph as part of an assignment, your instructor may want you to include an open circle for endpoints not included and a closed circle for included endpoints. Most textbook graphics use this convention. See page 74 of *Functions Modeling Change*.

Use TABLE with an initial value of 2 and an increment of 0.001. Then up arrow to see some values before and after 2. We see that the value of the function at 2 is 3, $g(2) = 3$. Therefore, the line $y = x + 1$ should have a closed circle as an endpoint at $x = 2$. When x-values are above 2 the function value is 1, so the horizontal graph $y = 1$ should have an open circle for the endpoint at $x = 2$.

2.4 Composite and Inverse Functions

[80]   Let $f(x) = x^2$ and $g(x) = 2x + 1$.
(a) Calculate $f(g(3))$ and $g(f(3))$.
(b) Find formulas for $f(g(x))$ and $g(f(x))$.

We enter the functions f and g as Y1 and Y2, and then use these to define the compositions as Y3 and Y4. Since we only want values of the compositions we deselect Y1 and Y2.

Use 2ndTBLSET to start at 0 with ΔTbl=1. The table shows the values $f(g(3)) = 49$ and $g(f(3)) = 19$. The composite functions are not equal to one another.

For part (b), the TI-84/83 cannot give an algebraic formula, but you can check the graph of your algebraic derivation against the graph of Y3 and Y4.

We graph in a ZDecimal window. Algebraically, we derive: $f(g(x)) = (2x + 1)^2$ and $g(f(x)) = 2x^2 + 1$. These algebraic answers are entered as Y5 and Y6 and graphed to check that they are the same as Y3 and Y4.

Tip: You can shorten a composition definition by not showing the variable X. For example, Y3=Y1(Y2), but showing X is closer to function composition symbolism.

Inverse Functions

The TI does not find the algebraic inverse of a function. However, using Solver allows you to easily find values of a function's inverse.

[81]   There is a special mathematical notation for the inverse value of a function. For example, the cricket chirp rate formula and its inverse would be written:

$$\text{if } T = f(R), \text{ then } R = f^{-1}(T).$$

This inverse notation cannot be used on the TI-84; however, Solver makes it somewhat unnecessary. We now see how to evaluate $f^{-1}(T)$.

Tip: Solver can have many variables, and it solves for any *one* variable using the values set for all the other variables.

Solver does inverses

In Solver the original function formula is entered as a difference $0 = T - f(R)$. Notice that we did not need to translate to X and Y variables; however, we did need to use the ALPHA key to enter T and R.

```
EQUATION SOLVER
eqn:0=T-(.25R+40
)■
```

After pressing ENTER, all the variables are listed. We now fill in the ones we know (in this example T=50), and position the cursor on the line of the variable that we want to know (in this example R). To solve, press ALPHA SOLVE.

```
T-(.25R+40)=0
 T=50
R=■
bound={-1ε99,1…
```

We see R=40. If $T = f(R)$, then Solver is evaluating $R = f^{-1}(T)$. In mathematical notation $f^{-1}(50) = 40$. In words this means that at 50 degrees the chirp rate is 40.

```
T-(.25R+40)=0
 T=50
■R=40
 bound={-1ε99,1…
■left-rt=0
```

The beauty of this system is that we need not get lost in a question about what is the function and what is the inverse function. With the same setup in Solver we can calculate either variable in terms of the other. We now calculate the temperature when the chirp rate is 50. Enter R=50.

```
T-(.25R+40)=0
 T=50
R=50■
bound={-1ε99,1…
left-rt=0
```

Arrow up to the line T=50 and press ALPHA SOLVE. We see the newly calculated value T=52.5. (The old value, T=50, was used as the guess for this calculation.)

In mathematical notation this result is written as $f(50) = 52.5$, and it means that at the chirp rate of 50, the temperature is 52.5 degrees.

```
T-(.25R+40)=0
■T=52.5
 R=50
 bound={-1ε99,1…
■left-rt=0
```

2.5 Concavity

Concavity and Rate of Change

For a linear function, the rate of change is constant and its graph has no concavity. If the function's rate of change is increasing we see the graph as concave up; if it is decreasing the graph is concave down. In general, concavity is often determined from a graph. When the concavity of a graph is hard to see, numerical methods are a better to determine concavity.

[84]   The carbon-14 decay function, $Q = 200(0.886)^t$, is not a linear functions; we use a list of rounded values for 0, 5, 10, and 15 thousand years to look at $\Delta Q / \Delta t$.

Enter Y₁=200(.886)^X and setting the viewing window to $0 \leq x \leq 20$ (Xscl=5) and $-50 \leq y \leq 250$ (Yscl=50), press the TRACE key to see the screen shown. This is clearly concave up.

We make a list L1 of the Q function values (rounded) for $t = 0, 5, 10, 15$. We then find $\Delta Q/\Delta t$ for the three consecutive intervals with $\Delta t = 5$. Since Δt is constant, we do not need to make a difference list for the t values, we just divided by 5. The rate values -18.2, -9.8, and -5.4 are increasing, that is, getting less negative. An increasing rate of change confirms upward concavity.

Tip: The value of the function is decreasing, but the value of the rate of change is increasing.

2.6 Quadratic Functions

The standard form of a quadratic equation is $y = ax^2 + bx + c$, where a, b, c are constants, $a \neq 0$.

Finding Zeros of a Quadratic Function

[89]   Find the zeros of $f(x) = x^2 - x - 6$.

Using a graph to find zeros

Using Y₁=X²-X-6 and Zstandard. We see the graph crosses the x-axis at, or close to, -2 and 3. Substitution confirms that these are the exact zeros.

Using Solver to find zeros

Use Solver and enter the standard quadratic equation. Then enter the specific coefficient of the equation and set a guess of X=0. This gives us our -2 answer.

Using the same equation and coefficients, we change the guess to some number X>1, and we find the other answer of 3.

```
A*X²+B*X+C=0
 A=1
■X=3
 B=-1
 C=-6
 bound={-1E99,1…
■left-rt=0
```

The quadratic formula to find zeros

[89]   The Solver method of solution starts with a guess and uses successive calculations to get closer and closer to the correct answer. This gives a very good approximate answer and in some cases an exact answer. The standard quadratic equation has a pair of zeros (solutions) that are written directly in the exact form:

$$x = \frac{-b \pm \sqrt{b^2 - 4ac}}{2a}.$$

This is not an approximation!

Since the equation is in standard form, we assign (use STO>) the coefficients to letter names. The colon (:) allows us to make all three assignments as a single command. (The colon symbol is above the decimal point key in ALPHA mode.) When using multiple assignments only the final one, here –6, is shown as output.

```
1→A: -1→B: -6→C
                -6
■
```

Now enter the famous quadratic formula and calculate the first zero of the function. Use 2ndENTRY to find the second zero by bringing back the entry and changing the plus sign in front of the square root to a minus sign. For new quadratics use deep recall to go back and change the A, B, C line and then reuse the two solution expressions shown here.

```
(-B+√(B²-4A*C))/
(2A)
                3
(-B-√(B²-4A*C))/
(2A)
               -2
```

The method above, which stores the quadratic formula, is hazardous in that after many non-quadratic formula calculations the formula may "fall off" the deep recall storage. The following optional technique is suggested for students who sleep better at night knowing that their quadratic formula is safely stored. You can write and store the quadratic formula in a simple program.

Optional: Programming Adventure

Press PRGM and arrow to NEW.

```
EXEC EDIT NEW
1■Create New
```

Name the program. Note that the ALPHA lock mode is automatically turned on.

```
PROGRAM
Name=QF
```

Pressing ENTER shows you a program edit screen. Each line automatically begins with a colon

```
PROGRAM:QF
:█
```

We want our program first to ask for the value of A, B, and C. Press PRGM and now, because you are already in program edit mode, it shows a menu screen. Arrow to I/O (Input/Output) and select 2:Prompt. This pastes that word on the program edit screen. (Commands pasted from menus are distinguishable by their lower case letters.)

```
CTL I/O EXEC
1:Input
2:Prompt
3:Disp
4:DispGraph
5:DispTable
6:Output(
7↓getKey
```

Finish this line by pressing the keys:

ALPHA A , ALPHA B , ALPHA C ENTER.

```
PROGRAM:QF
:Prompt A,B,C
:█
```

Complete the program as shown. The other lines display the formula for the two calculated zeros. They start with Disp from the PRGM I/O menu. Press 2ndQUIT to exit the program edit screen.

Note: parentheses are needed for correct division, there is a difference between subtraction and negation, and an explicit multiplication sign is needed between A and C.

```
PROGRAM:QF
:Prompt A,B,C
:Disp (-B+√(B²-4
A*C))/(2A)
:Disp (-B-√(B²-4
A*C))/(2A)
:
:
```

Tip: Programs can be much more complex. The *TI–Guidebook* has a complete chapter that more fully describes the programming capabilities. We started with this example because of its simplicity.

Using a program

From a cleared home screen, press PRGM. From the EXEC (Execute) menu, select the program you want to use. With the program name on the home screen press ENTER. Screens vary depending upon whether other programs have been entered.

```
EXEC EDIT NEW
1:FOCALE
2:QF
3:VECTORS
```

The program name appears on the top line, and A=? prompts you to enter A. Enter A, then B, then C. Notice here that the negation key is used to enter a negative number. The subtraction key causes an error. If an error occurs in this program, it is most likely a misuse of the negation and subtraction keys. The programmed formulas contain both negation and subtraction symbols, and they cannot be interchanged.

```
prgmQF
A=?1
B=?-1
C=?-6█
```

With luck you were able to see a screen with answers exactly as shown here. If not, go back (PRGM EDIT) and edit the program to correct the formulas.

```
prgmQF
A=?1
B=?-1
C=?-6
                3
               -2
            Done
█
```

Another example

[89]   Find the zeros of $h(x) = -\dfrac{1}{2}x^2 - 2$.

Use the program and enter A=-1/2, B=0, and C=-2.

```
A=?-1/2
B=?0
C=?-2
```

A simple graph of $h(x) = -\dfrac{1}{2}x^2 - 2$ shows that it does not cross the x-axis. Since there are no real solutions, the result is the error message: NONREAL ANS. The program can be changed to check for non-real solutions and avoid the error message. In Chapter 7 we address non-real solutions.

```
ERR:NONREAL ANS
1█Quit
2:Goto
```

CHAPTER THREE

EXPONENTIAL FUNCTIONS

3.1 The Family of Exponential Functions

Definition: An *exponential* function has a constant percentage rate of change. In this chapter we look at what this means in terms of tables, graphs, and formulas.

Growing at a Constant Percent Rate

[106] Suppose you are offered a job at a starting salary of $40,000 per year and the company promises annual raises of 6% per year for the first five years after you are hired. Compute the salary for the first few years.

The basic computation each year is:

$$\text{Next salary} = \text{Former salary} + 6\% \text{ of former salary.}$$

Using the TI sequence mode

The TI has a special sequence mode (Seq) that allows you to enter the initial salary and a formula for finding the next value. We now see how this works.

Press MODE and change the function mode from Func to Seq in the fourth row of the screen.

```
NORMAL  SCI   ENG
FLOAT   0123456789
RADIAN  DEGREE
FUNC  PAR  POL  SEQ
CONNECTED  DOT
SEQUENTIAL  SIMUL
REAL  a+bi  re^θi
FULL  HORIZ  G-T
SET CLOCK 01/28/07 3:00PM
```

The Y= screen now shows the definitions for three sequence defined functions named u, v, w. The first setting nMin= is preset to zero and defines our starting n. The u(n) is the function definition and u(nMin) is the initial value. In the next frame we see how our sequence function is defined.

```
Plot1 Plot2 Plot3
 nMin=0
\u(n)Bu(n-1)+.06
u(n-1)
 u(nMin)B{40000}
\v(n)=
 v(nMin)=
\w(n)=
```

Entering the formula requires that you use the 2nd u key (above the 7 key). Changing the function mode to Seq redefines the $\boxed{X,T,\theta,n}$ key to be the variable n, which is needed in the definition. The u(nMin) can be entered as 40000 without set brackets, but they are automatically inserted. (A more advanced feature allows this value to be a list. See the *TI–Guidebook*, if interested.)

```
Plot1 Plot2 Plot3
 nMin=0
\u(n)Bu(n-1)+.06

 u(nMin)B{40000}
\v(n)=
 v(nMin)=
\w(n)=
```

Before pressing the 2ndTABLE key to obtain the shown screen, use 2ndTBLSET to set TblStart=0 and ΔTbl=1. This table can be scrolled.

Growth Factor

The above function definition for u(n) could also have been written as u(n)=1.06u(n-1). In this form, 1.06 is called the growth factor. Notice that the function, u(n), is only defined for input values 0, 1, 2, and so on. It is called discrete. In truth, the salary function is a piecewise defined function that remains constant throughout the year and then jumps by 6%. We now model the salary function by a formula that has a domain of all real numbers but does not have a sudden jump.

Change the function mode back to Func.

Define Y1=40000(1.06)^X.

Set the viewing WINDOW to $0 \le x \le 20$ (Xscl=5) and $0 \le y \le 120000$ (Yscl=40000).

Press TRACE to show the equation and graph.

Use 2ndTBLSET to set TblStart=0 and ΔTbl=1. This table shows that the functions Y1 and u have the same value for 0, 1, 2, 3, …. The Y1 definition has the mathematical advantage that it is defined for all real numbers and is continuous (without jumps).

3.2 Comparing Exponential and Linear Functions

We now compare the tables and graphs of exponential and linear functions.

Identifying Linear and Exponential Functions from a Table

[115] The table below gives values of two functions, *f* and *g*. Which function is linear and which is exponential?

x	20	25	30	35	40	45
f(*x*)	30	45	60	75	90	105
g(*x*)	1000	1200	1440	1728	2073.6	2488.32

In this example Δx is constant and it is easy to see that the difference between consecutive values of $f(x)$ is 15. (We could use ᴧList as in Section 2.1 to identify linear functions.) The difference between consecutive values of $g(x)$ is not constant, but the ratio is constant. We identify $g(x)$ as an exponential function by inspecting the ratios.

Using lists to find quotients

To find consecutive ratios we make two lists: one is the numerator and one is the denominator. We put all but the last entry for $g(x)$ into list L1. (Recall that set brackets and list names require the 2nd key and that the arrow ($\rightarrow$) is the STO> key.) Next edit this list by deleting the first entry, adding the last entry, and storing it as L2.

Now L2/L1 is the list of consecutive ratios. We see that the ratios are all 1.2, which indicates that $g(x)$ is an exponential function. It should be noted that this method of ratio checking requires that Δx be a constant.

> *Tip:* Some students notice that tables with decimal values are likely to be exponential functions, while linear functions are more likely to have whole numbers. While often true, it is not always true; be careful about assuming too much from a few table values.

Exponential Growth Always Outpaces Linear Growth

[118] Thomas Malthus made gloomy predictions about long-run food supply. He assumed that the population grew exponentially and that food supply grew linearly. For example, if P is the population (in millions) and N is the number of people (in millions) fed from the food supply then we can investigate this model with the two equations

$$P = 2(1.04)^t \text{ and } N = 4 + 0.5t.$$

Enter these functions as Y1 and Y2. Set the viewing WINDOW to $0 \le x \le 100$ (Xscl=10) and $0 \le y \le 100$ (Yscl=10). TRACE shows the graph. We see that the linear equation outpaces the exponential equation at the 50-year mark, but later the exponential function passes the linear function.

The intersection point is when food shortage begins. To find this point, you can use the 2ndCALC 5:intersect feature introduced in Section 1.5. (The intermediate steps are not shown.) We find that the exponential function is greater than the linear function after about 78 years.

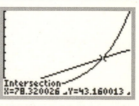

3.3 Graphs of Exponential Functions

In this section we look at some special features of an exponential graph.

The Exponential Family

[123]   Our basic exponential functions have the form $Q = ab^t$, where a and b are parameters. You can graph a related group of functions to see the effect of changing a parameter. Recall that while it is possible to write each function as its own Y= formula, the set notation is used to simplify the process. Although the screens are not shown here, you should separately graph each of the following:

$$Y_1=\{150,100,50\}*1.2^\text{X} \qquad \text{and} \qquad Y_2=500\{1.4,1.2,.8,.6\}^\text{X}$$

Horizontal Asymptotes

[124]   The horizontal asymptote of a function is a line $y = k$, such that as $x \to \infty$ or $x \to -\infty$ then $f(x) \to k$. To check this numerically, you can use large values of x in a table, or graph with a very wide window.

Exponential functions of the form $Q = ab^t$ have a horizontal asymptote of $y = 0$.

After entering Y1=200(.9)^X and setting the viewing window to $0 \le x \le$ 60 (Xscl=15) and $-50 \le y \le 250$ (Yscl=50), press the TRACE key to see the screen shown. The graph falls toward the x-axis, and at the far right gets so close to the axis that the axis and curve share the same screen pixel points. Trace over to see that the function values are getting closer and closer to zero.

Set ▵Tbl=30 and look at the table. (The E notation means 6.5E-4 = $6.5*10^{-4} = 0.00065$.) Thus, the function values are getting closer and closer to zero.

Solving Exponential Equations Graphically

In Section 2.4 we used Solver to give us a numerical solution; now we take a graphical approach.

[125]   Suppose that a 200 µg sample of carbon-14 decays according to the formula,

$$Q = 200(0.886)^t,$$

where t is in thousands of years. Estimate when there will be 25 µg of carbon-14 left.

Graph Y1=200(.886)^X and Y2=25. Set the viewing WINDOW to $0 \le x \le 30$ (Xscl=5) and $-10 \le y \le 50$ (Yscl=10). Using TRACE, you can arrow to the intersection and estimate $x \approx 17.2$.

Alternatively, you can use the 2ndCALC intersect feature to get a more accurate estimate as shown here.

Fitting Exponential Functions to Data

[126]   The population of the Houston Metro area has been growing exponentially. If t is in years since 1900, the table below shows the population in thousands.

t, years since 1900	0	10	20	30	40	50	60	70	80	90	100
P, thousands	184	236	332	528	737	1070	1583	2183	3122	3733	4672

The TI can derive an exponential regression formula in a simple sequence of steps.

Exponential Regression

Enter the data using STAT Edit.

Use 2ndSTATPLOT to turn on the Plot1 graph and select the lists. The lists may be already selected as Y₁ and Y₂.

Use ZoomStat to see a data plot. It is clear by the curve of the graph that a linear regression equation does not seem appropriate.

On a fresh home screen, use STAT CALC and select ExpReg (Exponential Regression) to begin the command. This process is the same as the one we used for linear regression in Section 1.6.

After pasting ExpReg, specify the lists to use for the regression and a Y= equation for the formula to be pasted into.

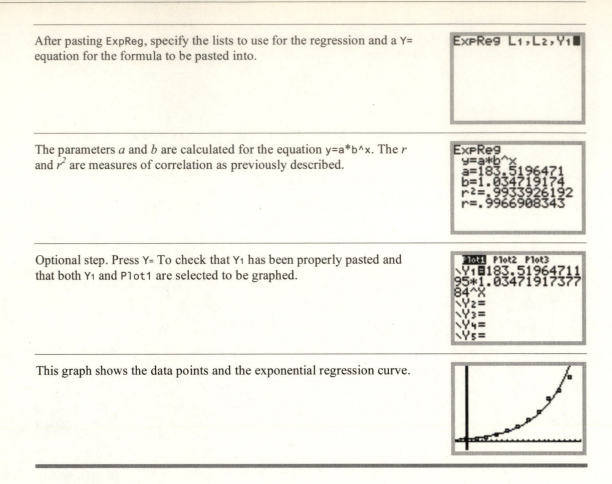

The parameters *a* and *b* are calculated for the equation y=a*b^x. The *r* and r^2 are measures of correlation as previously described.

Optional step. Press Y= To check that Y1 has been properly pasted and that both Y1 and Plot1 are selected to be graphed.

This graph shows the data points and the exponential regression curve.

3.4 Continuous Growth and the Number e

[130]   The number *e* is a constant, rivaled only by the more famous constant pi. One of the derivations of *e* is found by using the formula for periodic interest payment. The annual growth rate for an investment with interest rate 100% that is paid *n* times a year is

$$\left(1 + \frac{1}{n}\right)^n$$

Enter the formula in Y1 and use the TABLE to display the first 7 values. This screen shows that if you arrow up to highlight Y1 in the header, its formula is displayed. The more common *n*, here represented by X, are quarterly, monthly, yearly, etc.. We see how to make a table of special values in the next frame.

Press 2ndTBLSET to see the setup menu. Change Indpnt: to Ask, by first highlighting it and then pressing ENTER. Press 2ndTABLE and you see an empty table screen, ready for custom X entries.

Enter the values of X that are of interest (no pun intended.) The value $n = 365*24 = 8760$ is compounded hourly; the value $n = 365*24*60 = 525600$ is compounded every minute. Note that by highlighting a Y1 cell value, the complete decimal accuracy is shown on the bottom line.

Use 2ndQuit to return to the home screen. It is also possible to directly evaluate special values of Y1. The value shown for $n = 31536000$ represents compounding every second. To check that we are indeed getting closer and closer to e, we display its value by entering 2nde.

There are two e keys on the TI. The 2nde key, on the right side of the keypad, is used in the above example. The ex key, on the left side of the keypad, is use when function notation is needed.

[132] For example, after 8 hours, the amount of caffeine left in the body from drinking a cup of coffee with 100 mg of caffeine is given by

$$A = 100e^{-0.17(8)}.$$

The 2ndex key allows easy evaluation of the function $f(t) = ae^{kt}$.

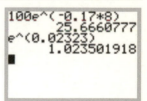

Use the key 2ndex to paste the expression e^(in the caffeine formula. Another example, from page 133, uses the value of e^(0.02323) to find the equivalent of a 2.323% continuous yearly rate to be a 2.35% annual rate.

3.5 Compound Interest

Applications involving exponential equations are commonly written with base e formulas. We need to be able to convert an exponential equation to an equivalent form in another base. So why is this e base so important? It models growth that has a continuous growth rate. This is in contrast to a periodic growth rate such as an annual growth rate.

Compound Interest Formulas

Growth rates are commonly used in financial applications. We often need to compare rates that are in different units. Annual interest rates are often advertised as percentages in clean decimals; these are called nominal rates. When a nominal rate is compounded more frequently than once a year, then an effective rate is determined. Effective rates often have long strings of decimal digits and need to be rounded.

Converting Nominal and Effective Rates

[136]   What are the effective annual rates of an account paying 12% interest, compounded annually?, compounded monthly? What is the effective rate of an account paying 6% annual interest compounded daily?

The TI has two conversion commands built in:

$$\blacktriangleright \texttt{Eff}(\textit{nominal rate, compounding periods})$$

$$\blacktriangleright \texttt{Nom}(\textit{effective rate, compounding periods}) .$$

The ▸ Eff(command is available in the catalog. Pressing CATALOG E moves to the items with first letter e. Likewise, CATALOG N moves close to ▸ Nom(.

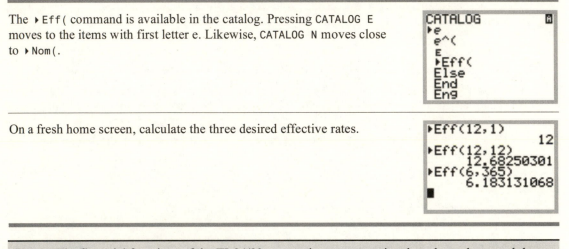

On a fresh home screen, calculate the three desired effective rates.

> *Tip:*   The financial functions of the TI-84/83 are much more extensive than shown here, and they are quite useful in a business finance course. The interested reader should see Chapter 14 in the *TI–Guidebook*.

---

[137]   The underlying formula relating nominal and effective rates is given as

$$B = P \cdot \left(1 + \frac{r}{n}\right)^{nt} ,$$

where r is the annual nominal rate and n is the number of times compounding is done in a year. Time, t, is in years and $\$P$ is the initial deposit.

Letting P be \$1 and letting t be 1 year, for a given r and n we can find the annual growth factor. This growth factor equals 1 + effective rate, so that we can easily read off the effective rate from the growth factor.

In the first calculation we find the growth factor of compounding daily with a nominal rate of 6%. Subtract 1 from the growth factor to see that this is the same effective rate shown at the bottom of the previous screen. The second calculation says that the effective rate of hourly compounding of a 6% nominal rate is 6.1836329%.

Continuous Compounding Interest

As we saw in Section 3.4, by compounding as ridiculously often as hourly (or every second), the effective rate reaches a limit called the *continuous* rate. It is given as

$$B = Pe^{rt}$$

where r is the annual nominal rate and time, t, is in years, and P is the initial deposit.

Once again for a given r, if we let P be \$1 and t be 1 year, we can find the annual growth factor. This growth factor is 1+effective rate, so we can easily read off the effective rate for a continuously compounded rate. This means that we can compare rates of all types.

[138] Which is better: An account that pays 8% annual interest compounded quarterly or an account that pays 7.95% annual interest compounded continuously?

We compare the growth factors to see that the lower nominal rate compounded continuously is slightly better. The difference would only become significant with large deposits or deposits held for many years.

Tip:	Be on the lookout for the word *continuous* in problems with rates. This is the tip-off that you should be using the formula in base e.

Notes:

CHAPTER FOUR

LOGARITHMIC FUNCTIONS

4.1 Logarithms and Their Properties

Logarithms Are Exponents

[152] We start by confirming that logs are exponents. For example, $\log 100 = 2$ because 2 is the exponent of 10 that gives 100, or $10^2 = 100$.

Here we confirm $10^{\wedge}\log(x) = x$ for $x = 100$. Note the use of Ans. In this case it might have been faster to type 10^2, but in most cases the log value is a decimal and cumbersome to re-enter.

```
log(100)
                  2
10^2
                100
■
```

Now a little TI magic trick. Functions such as the log allow a list as input. We first use {100, .01, 30} as input; the output is a list of function values. Now enter 10^Ans. Voila! The original list returns.

```
log({100,.01,30}
)
{2 -2 1.4771212…
10^Ans
        {100 .01 30}
```

Solving with Logs

If we want to find a value of Q in $Q = ab^t$ for a given value of t, we say we are *evaluating* Q at t. But when we want a value of t for a given value of Q, we say we are *solving* for t. To solve for t requires that we undo an exponent. This is what the $\log(x)$ and $\ln(x)$ functions do. It is advisable that you be able to solve both exponential and logarithmic functions by using algebra. Solving algebraically requires some decision about whether to use log() or ln() in finding the exact form answer. To find numerical approximations, Solver requires no decision about the kind of log function to use.

Solving an exponential equation where we normally use log

[154] Solve $100 \cdot 2^t = 337,000,000$, for t. To solve algebraically we would normally use the log function, but by using Solver there is no need to worry about which logarithmic function to use. However we must be satisfied with an approximation, which is a non-exact answer.

Enter the equation 0=100*2^X-337000000 in `Solver` and press ALPHA
SOLVE with the cursor on the X= line.

```
100*2^X-33700…=0
■X=21.684317160…
 bound={-1ᴇ99,1…
■left-rt=0
```

Solving an exponential equation where we normally use ln

[155] Solve $5e^{2x} = 50$ for x. Here the ln() function would normally be used to solve algebraically, but the
`Solver` command approximates the solution in exactly the same way as the previous example.

Arrow up to the old Solver equation and enter the new equation:
0=5e^(2X)-50. Press ALPHA SOLVE with the cursor on the X= line.

```
5e^(2X)-50=0
■X=1.1512925464…
 bound={-1ᴇ99,1…
■left-rt=0
```

The TI is log friendly

[156] Find the value of $\log\dfrac{17}{3}$. On some calculators pressing the following five key sequence: log 1 7 / 3
results in the value 0.410, which is incorrect. The TI is foolproof in this regard because it automatically enters a left-parenthesis after log.

First press the five key sequence: log 1 7 ÷ 3 and then ENTER. The TI
automatically inserts a parenthesis after the log. It allows you to not
close the parenthesis before pressing ENTER. The last calculation shows
how some calculators would interpret the five-key sequence and arrive at
0.410.

```
log(17/3
          .7533276667
log(17/3)
          .7533276667
log(17)/3
          .4101496405
```

4.2 Logarithms and Exponential Models

There are many applications of logarithms, but their main use is to solve exponential equations. In
particular, we show how to find doubling time and half-life for exponential models. We conclude with
the important note that logarithms cannot solve all types of exponential equations.

Models

[159] The US population, P, in millions, is currently growing according to the formula:

$$P = 299e^{0.009t},$$

where t is in years since 2006. When is the population predicted to be 350 million?

Press MATH 0:Solver, arrow up, and enter the equation shown after the eqn:0= prompt. Notice that the equation is entered as a difference that is set to zero.

```
EQUATION SOLVER
eqn:0=P-299e^(0.
009T)
```

After pressing ENTER we set the P to 350 and arrow down so that the cursor is on the T variable line. To solve for T, press ALPHA SOLVE. The square dot to the left of T indicates that it has been calculated. The US population is projected to be 350 million 17.5 years from 2006.

```
P-299e^(0.009…=0
 P=350
■T=17.498842343…
 bound={-1ɛ99,1…
■left-rt=0
```

Doubling Time

[161]   Find the time needed for a turtle population described by the function $P = 175(1.145)^t$ to double its initial size.

For 175 (the initial value) to double, we are solving $2 \cdot 175 = 175(1.145)^t$ for t. However, after a division by the initial value, this is the same as solving $2 = (1.145)^t$. Similarly for any exponential model, $Q = ab^t$, the doubling time is found by solving $2 = b^t$. The solution is $t = \dfrac{\log(2)}{\log(b)}$ and can now be found with your calculator.

On this screen we see that the solution to the turtle doubling time is independent of which logarithmic function we use. Our turtle population doubles roughly every five years.

```
log(2)/log(1.145
)
        5.119080084
ln(2)/ln(1.145)
        5.119080084
■
```

Half-life

[162]   While doubling time is found for an increasing exponential function, half-life is commonly found for an exponentially decaying function. For example, carbon-14 decays at an annual rate of 0.0121%. Find the half-life of carbon-14.

In general, half-life is found by solving $\frac{1}{2} = b^t$ for t. We note that the growth rate is $b = 1 - 0.000121$, and we solve using the log (or ln) function. The half-life of carbon-14 is 5728 years. (See the tip below if you are confused by 0.0121% = 0.000121.)

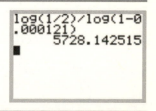

```
log(1/2)/log(1-0
.000121)
        5728.142515
■
```

| *Tip:* | Use care when entering percentages; they must be entered as their decimal equivalent. For example, 5% = 0.05, and for a fractional rate, 0.5% = 0.005. |

> *Tip:*   It is safer to enter expressions without simplification. They are easier to double check later for correctness.

Applications involving exponential equations are commonly written with base e formulas. We need to be able to convert an exponential equation to an equivalent form in another base. So why is this e base so important? It models growth that has a continuous growth rate. This is in contrast to a periodic growth rate such as an annual growth rate.

Converting bases

[163]   Convert the function $P = 175(1.145)^t$ to the form $P = ae^{kt}$. Convert the function $P = 7e^{0.3t}$ to the form $P = ab^t$.

   In general, the equality of the two forms means $ab^t = ae^{kt}$. Since a is the same in both cases, the equality becomes $b^t = e^{kt}$. With t also being the same, we further simplify to see that we need only solve $b = e^k$. In the first example the solution of $1.145 = e^k$ is $k = \ln(1.145)$. In the second we have $b = e^{0.3}$. Seeing that the conversion is finding k from b or b from k means a simple calculation on the home screen. However, we show off Solver as a single technique that converts either way.

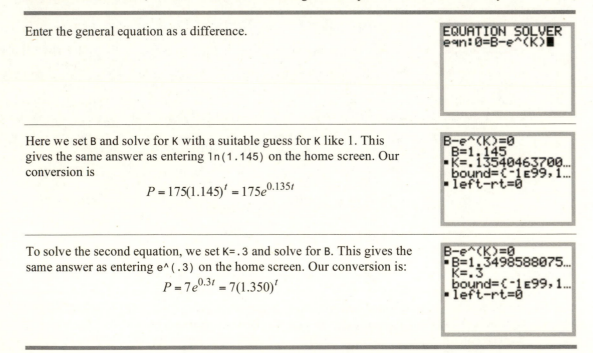

Enter the general equation as a difference.

```
EQUATION SOLVER
eqn:0=B-e^(K)█
```

Here we set B and solve for K with a suitable guess for K like 1. This gives the same answer as entering ln(1.145) on the home screen. Our conversion is

$$P = 175(1.145)^t = 175e^{0.135t}$$

```
B-e^(K)=0
 B=1.145
•K=.13540463700…
 bound={-1ε99,1…
•left-rt=0
```

To solve the second equation, we set K=.3 and solve for B. This gives the same answer as entering e^(.3) on the home screen. Our conversion is:

$$P = 7e^{0.3t} = 7(1.350)^t$$

```
B-e^(K)=0
•B=1.3498588075…
 K=.3
 bound={-1ε99,1…
•left-rt=0
```

Not All Exponential Equations Can Be Solved by Logarithms

[163]   With t in years, the population (in millions) of a country is given by $P = 2(1.02)^t$, while food supply (in millions of people who can be fed) is given by $N = 4 + 0.5t$. Determine the year in which the country first experiences food shortages.

   If we try to algebraically solve $4 + 0.5t = 2(1.02)^t$ using logs, we cannot isolate t; however, there is a numerical solution. (In fact, there are two such solutions.) If we blindly use solve, we may find an unacceptable solution; the relevant solution is the 199th year.

Use `Solver` with the difference of the two formulas:

$$0 = 4 + .5T - 2(1.02)^T$$

Warning: There are two solutions to this equation. If the T guess is not 129 or greater than the second solution T=-4.329 is shown.

```
4+0.5T-2(1.02..=0
■T=199.38071210…
 bound={-1ε99,1…
■left-rt=0
```

> *Tip:*   In the previous example, it would have been safer to use a graphical approach, since you would then see where the graphs intersect and be assured that you have a relevant solution. Safety has its price; a graphical approach has more steps.

The difference between log and ln

Why are there two logarithm keys? The answer is that `log` undoes exponents to the power of 10, while the other `ln` undoes exponents to the power of e. Either one can be used in solving any exponential equation. However, the ln is simpler for exponential equations involving e, and the log is preferred for equations with powers of 10.

Consider solving the first equation of this section, $350 = 299e^{0.009t}$. You can verify using algebra and logarithm properties that a solution using ln is: $t = \ln(350/299)/0.009$. If you used log to undo the exponent, then the solution is written $t = \log(350/299)/(0.009\log(e))$. The screen shows that these are the same solution.

```
ln(350/299)/0.00
9
          17.49884234
log(350/299)/(0.
009log(e))
          17.49884234
■
```

> *Tip:*   The value of e can be pasted from the keypad with 2ⁿᵈe (above divide) or pasted from the CATALOG.

4.3 The Logarithmic Function

[167]   The graph of $\ln(x)$ is shown in this example.

This screen shows the graph of Y1=ln(X) using the ZDecimal window. When TRACE is used we see that for X=0 there is no Y value. This means that the function is undefined at this point. By tracing to the left, you see that it is also undefined for negative values of x.

Trace to the right and Y values are shown. In this case we see that the ln function has a zero (also called a root) at $x = 1$. The property is listed as $\ln 1 = 0$.

How can we show the property $\ln(e) = 1$? If you trace by arrow keys and attempt to find X when Y=1, you must settle for an approximate result such as that shown on the bottom line. It says: $\ln(\approx e) = \approx 1$. In trace mode, you can set X=e and see that $\ln(e) = 1$.

The Graph of ln(*x*) and *e*ˣ

[170]   The two functions $y = \ln(x)$ and $y = e^x$ are inverses. Their graphs are reflections of one another across the line $y = x$.

Enter three functions: Y1=ln(X), Y2=e^(X), and Y3=X. Graph with a ZDecimal window.

> *Tip:*   ZDecimal was used in the previous example because it shows the $y = x$ line at a true 45° angle. If you use a ZStandard window, the graphs do not look like perfect reflections. But you can use ZOOM 5:ZSquare to square up any given window.

Table Values for ln(*x*)

What can we expect to see from a table of ln values?

Here we proceed showing both Y1 and Y2 from the previous example. We know the ln function is undefined for $x = 0$, and this is denoted as ERROR. For this table the ΔTbl has been set to 1 and both variable settings are Auto. We note that both of these functions are increasing, with the ln function growing rather slowly.

To find self-selected values for a table, we use TBLSET and set Indpnt: Ask. Recall from Section 3.4 that this allows the independent variable x to be entered by hand. It ignores the settings for TblStart and ΔTbl.

Here we see a screen showing that we have manually requested the values for 0, 1, and 2. The bottom of the screen shows that we have just pressed 2ⁿᵈe. After pressing ENTER you will see that $\ln(e) = 1$.

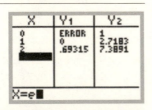

> *Tip:*   With Indpnt: set to Ask, you can mix the order or size of x table entries. Even special values like π and e can be used.

4.4 Logarithmic Scales

Log-Log Scales

[179]   Plot the table below, which shows the average metabolic rate in kilocalories per day for animals of different weights.

Animal	Weight(lbs	Rate(kcal/day)
Rat	1	35
Cat	8	166
Human	150	2000
Horse	1750	9470

Use STAT Edit to enter the data into two lists as shown.

Use 2ndSTATPLOT to set Plot1 as shown. (Don't forget to be sure that the Y= defined functions are turned off.)

Use ZoomStat to set a good window for the data.

This does not look like a very good plot because some of the values are squeezed at the origin.

The remedy is to change to a Log-Log scale. This means that we graph the data, with both *x*- and *y*-axes being in log scale. We convert L1 to log values and store them in L3. (Arrow up to highlight L3, and then enter the command shown on the bottom definition line.)

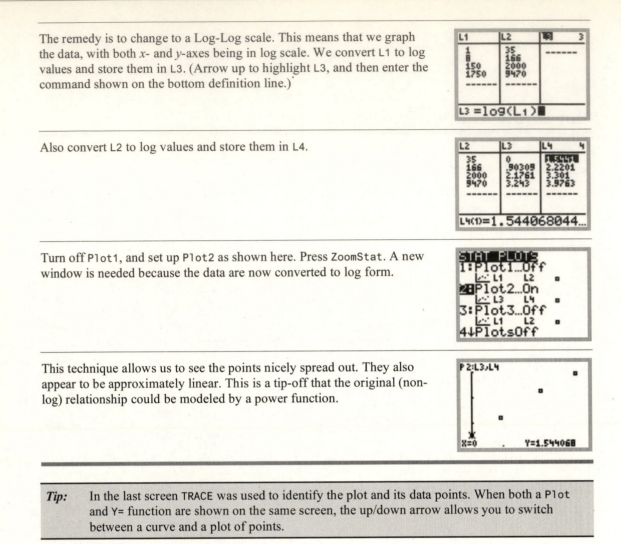

Also convert L2 to log values and store them in L4.

Turn off Plot1, and set up Plot2 as shown here. Press ZoomStat. A new window is needed because the data are now converted to log form.

This technique allows us to see the points nicely spread out. They also appear to be approximately linear. This is a tip-off that the original (non-log) relationship could be modeled by a power function.

Tip: In the last screen TRACE was used to identify the plot and its data points. When both a Plot and Y= function are shown on the same screen, the up/down arrow allows you to switch between a curve and a plot of points.

In Section 1.6 we saw how to fit a linear regression equation to data. In this section we find how to fit either an exponential regression equation to data.

The Sale of Compact Discs

[180]   The table below shows the fall in sales of vinyl long playing records (LPs) and the rise of compact discs (CDs) during the years 1982 through 1993. We use $t = 0$ for 1982.

t, years since 1982	c, CDs (millions)	l, LPs (millions)
0	0	244
1	0.8	210
2	5.8	205
3	23	167
4	53	125
5	102	107
6	150	72
7	207	35

8	287	12
9	333	4.8
10	408	2.3
11	495	1.2

The textbook's approach to fitting a curve to this data is to use the log function to *linearize* the data. This yields a linear regression equation for the converted data that can then be used to algebraically derive the exponential regression formula. This approach is good for conceptual understanding, and your instructor may want you to do the work that way. However, the TI can derive the exponential regression formula in a simple sequence of steps.

(Optional) Exponential regression: linearized by a log scale

Enter the data using STAT Edit. Arrow to the header of L3. Press ENTER to be prompted for the formula shown at the bottom of the screen. In the screen shown there are five more lines of data that are not shown.

Now L3 has the ln values of each entry in the L2 list.

On a fresh home screen, use STAT CALC and select LinReg(a+bx) (Linear Regression) to begin the command. This process is the same as the one we used in Section 1.6.

After pasting LinReg(a+bx), specify the regression lists as the first and the third, L1 and L3. We are finding a linear regression on L1 and L3=ln(L2).

The parameters a and b are calculated for the equation y=a+bx. But this is really $\ln(y) = a + bx$ so that to reach an exponential regression equation we take both sides to the power of e.

We calculate $e^a \approx 250$ and round b to define:

$$Y_1 = 250e^{\wedge}(-.011X)$$

With both Y₁ and Plot1 selected we graph in a ZoomStat window.

```
e^(5.52)
        249.6350372
```

The graph shows the data points and the exponential regression curve.

```
Y1=250e^(-0.011X)




X=247.5      Y=16.427568
```

Exponential regression: Using built-in regression equation

Using only L1 and L2 of the previous example, we now apply the TI's built-in ExpReg to reach the same result. On a fresh home screen, use STAT CALC and select ExpReg

```
EDIT CALC TESTS
5↑QuadReg
6:CubicReg
7:QuartReg
8:LinReg(a+bx)
9:LnReg
0:ExpReg
A↓PwrReg
```

This specifies calculating an exponential regression using L1 and L2 and pasting the formula into Y₁.

```
ExpReg L1,L2,Y1■
```

We see the coefficients, but it is still unclear whether we have the same formula since the previous one was given in the form $y = ae^{kx}$ and this one is of the form $y = ab^x$.

```
ExpReg
 y=a*b^x
 a=249.3500705
 b=.9891424384
 r²=.9858422773
 r=-.9928959046
```

We see that the two forms of the equation are essentially the same.

```
e^(-0.011)
        .9890602788
■
```

Since Y₁ has been pasted as part of the ExpReg command, we can press GRAPH and see the same plot and curve as derived by the previous method.

CHAPTER FIVE

TRANSFORMATIONS OF FUNCTIONS

5.1 Vertical and Horizontal Shifts

The graph of a function may be shifted vertically or horizontally. In this section we look at the change in a function's formula that produces a shift of its graph.

Inside and Outside Changes

[197] If $n = f(A) = \dfrac{A}{250}$ gives the number of gallons of paint needed to cover a house of area A ft^2, show the difference between $f(A+10)$ and $f(A)+10$ numerically.

Enter Y₁ in the Y= menu. Now define Y₂ and Y₃ in terms of Y₁ as shown. Recall that the Y₁ used in the Y₂ definition must be pasted from the VARS Y-VARS 1:Function screen.

```
Plot1  Plot2  Plot3
\Y1∎X/250
\Y2∎Y1(X+10)
\Y3∎Y1(X)+10
\Y4=∎
\Y5=
\Y6=
\Y7=
```

Before pressing 2ⁿᵈTABLE, use 2ⁿᵈTBLSET to start the table at $x = 0$, and have it go up by 10 for each x. In the table shown, the cursor has been moved to exhibit the Y₂ formula. Looking at the table, we see that a value of Y₁ appears back 10 units (one line up) in the Y₂ table.

X	Y₁	Y₂
0	0	.04
10	.04	.08
20	.08	.12
30	.12	.16
40	.16	.2
50	.2	.24
60	.24	.28

Y₂∎Y1(X+10)

Return to the Y= screen and turn off Y₂. Now press 2ⁿᵈTABLE to compare Y₁ and Y₃ values. Here we see that each Y₃ value is just the Y₁ values plus 10.

X	Y₁	Y₃
0	0	10
10	.04	10.04
20	.08	10.08
30	.12	10.12
40	.16	10.16
50	.2	10.2
60	.24	10.24

Y₃∎Y1(X)+10

Tip: To evaluate a function for a single value, the Y₁(0) method is quite handy, but if you wanted to see many values, say from 0 to 12, then a table is the best choice.

Horizontal Shifts

Graph $f(x) = x^2$. Graph f shifted to the right two units.

Define the function f as Y₁. Shifting to the right means that the variable x is replaced by x–2. So the shifted graph is defined as Y2=Y₁(X-2).

Use a ZDecimal window to see the two graphs. Indeed the graph of Y2 is the graph of Y₁ shifted two units to the right. Although not used here, the TI graph style feature would allow you to draw Y2 as a thicker graph so that you can visually distinguish between the two graphs.

Vertical Shifts

Graph $f(x) = x^2$. Graph f shifted down one unit.

With the function f as Y₁, shifting down one unit means the value of the function Y₁ must be one unit less. So the shifted graph is defined as Y3=Y₁(X)-1. In this example we turned off the Y2 definition.

Using a ZDecimal window we see the two graphs. Indeed, Y3 is just Y₁ shifted one unit down.

Combining Shifts

[199]   With $f(x) = x^2$, graph f shifted to the right two units and down one unit.

Here we make a subtle change to the definition of Y3; it is defined in terms of Y2. Thus Y3 depends on Y2, and Y2 depends on Y₁. Note that even though Y2 is not selected to be graphed, it can still be used as part of another definition.

Using the ZDecimal window we see the two graphs. We see Y3 as Y1 shifted two units to the right and one unit down.

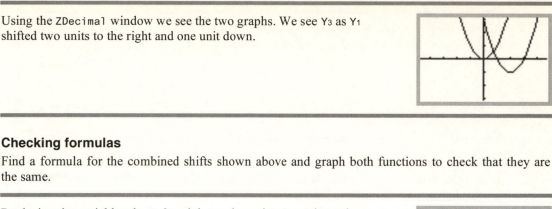

Checking formulas

Find a formula for the combined shifts shown above and graph both functions to check that they are the same.

Replacing the variable x by $x-2$ and then subtracting 1, we have the combined shift formula shown as Y4. Since we want to check that the two graphs Y3 and Y4 are the same, we need some way of knowing that the second function is graphing the same curve as the first. After all, it could just be graphing outside the window. We use the fifth graph style, which displays a circular cursor as it graphs.

After the first graph is drawn, the second graph is seen as the same because the circular cursor traces out the same points. (See the tip below if you want to make the cursor pause as in this example.)

Tip: Press ENTER to pause a graph and then ENTER again to resume graphing. The pause indicator is twinkling dots in the upper-right corner of the screen. Press ON to break and abort active graphing or calculating.

5.2 Reflections and Symmetry

A reflection over the x-axis corresponds to an outside change, while a reflection over the y-axis corresponds to an inside change. Symmetry occurs when a reflection gives the same graph.

Reflections

[203] The table below gives values of a function f. Write a formula for $f(x)$ and graph f.

x	-3	-2	-1	0	1	2	3
$f(x)$	1	2	4	8	16	32	64

In this example Δx is constant, and it is easy to see that the values of $f(x)$ are doubling. We identify $f(x)$ as an exponential function with growth factor 2. Also we see that the initial value (when $x = 0$) is 8. Thus our formula is $f(x) = 8(2)^x$.

Set the viewing window to $-3 \leq x \leq 3$ (Xscl=1) and $-65 \leq y \leq 65$ (Yscl=10) and graph f. It shows the graph is increasing and concave up.

To reflect the graph f across the x-axis, graph $g(x) = -f(x)$. Use Y2 in terms of Y1.

In this and the next two screens, we use the thick graph style to make the reflected graph show as a darker curve. Recall that this style is set in the Y= screen on the far left of the definitions.

To reflect the graph f across the y-axis, graph $h(x) = f(-x)$. Use Y3 in terms of Y1 and turn off Y2.

To reflect the graph f across the origin, graph $k(x) = -f(-x)$. Use Y4 in terms of Y1 and turn off Y3.

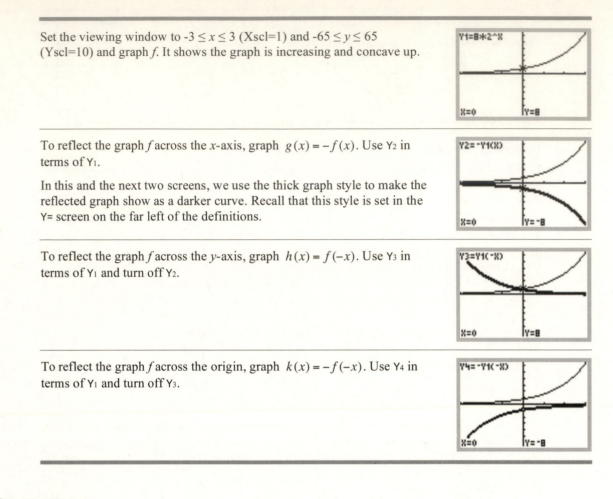

Symmetry

[207]   Determine whether the following functions are symmetric across the y-axis, the origin, or neither.

$$\text{(a) } f(x) = |x| \quad \text{(b) } g(x) = 1/x \quad \text{(c) } h(x) = -x^3 - 3x^2 + 2$$

Enter these functions as Y1, Y2, and Y3. Set the viewing window to ZDecimal for the graphing. We see the graph of $f(x)$ suggests the function is symmetric about the y-axis. Algebraically, we can verify that $f(-x) = f(x)$. You may also want to check by graphing Y4=Y1(-X) and comparing it to the graph of Y1.

The graph of $g(x)$ suggests that the function is symmetric about the origin. Algebraically we can verify that $-g(-x) = g(x)$. You may also want to check by graphing Y4=-Y2(-X).

The graph of $h(x)$ suggests that the function has no symmetry. Algebraically, we can verify that $h(-x) \neq h(x)$ and $-h(-x) \neq h(x)$.

We see that $h(-x) \neq h(x)$ by graphing Y4=Y3(-X) with Y4 in thick graph style. The two graphs are clearly not the same. Graph Y4=-Y3(-X) to see that h has neither symmetry.

5.3 Vertical Stretches and Compressions

A vertical stretch or compression is represented by an outside change to its formula.

Stretch Factor

[213]   A yam is placed in a 300°F oven. The table gives $r(t)$, the yam's temperature t minutes after being placed in the oven. Make a table of values for $q(t) = 1.5r(t)$. Graph the functions.

t	0	10	20	30	40	50	60
$r(t)$	0	150	225	263	281	291	295

Enter the data in L1 and L2; then create a new list L3=1.5*L2. To calculate this L3, arrow up to highlight the L3 header and press ENTER. The cursor appears on the bottom line to enter the calculation. Press ENTER again and the calculated numbers appear in L3.

Since we have lists without formulas, we must use STATPLOT to graph these values. Press 2ndSTATPLOT to see the current settings for the three possible plots. Turn on Plot1 and Plot2 and select the settings shown. (The intermediate plot setting screens are not shown here.) An xyLine plot has been selected; it is similar to a scatter plot, but connects consecutive data points with a line.

After the settings are made, use ZoomStat to graph the plots. We see that the second graph is a stretch of the first graph. If this graph has extra curves, you may have forgotten to turn off function graphing in the Y= list.

Average Rate of Change

[214]   How is the average rate of change of a function affected by a stretch? We show numerically that if $g(x) = k \cdot f(x)$ then on any interval,

$$\text{Average rate of change of } g = k \cdot (\text{Average rate of change of } f).$$

We start with the three lists from the previous example. In L4 we calculate the average rate of change for the original function. Since $\Delta x = 10$ for all the intervals, we divide by 10 and avoid using ∆List(L1).

In L5 we calculate the average rate of change for the stretched function.

If you are arithmetically swift, you see that L5=1.5*L4. To check this, you can use L6=1.5*L4. You find that L5 is the same as L6.

5.4 Horizontal Stretches and Compressions

[222]   Match functions $f(t) = e^t$, $g(t) = e^{0.5t}$, $h(t) = e^{0.8t}$, $j(t) = e^{2t}$ with the graphs.

Enter the four functions in the Y= menu. Turn off Plot1 and Plot2 if selected by using the up arrow.

Set the viewing window to $0 \le x \le 10$ (Xscl=1) and $-20 \le y \le 100$ (Yscl=50) and graph. Without using trace, see if you can identify which graph is which. Here TRACE has been turned on, and the cursor has been moved to identify the first curve.

The order of the functions from left to right is Y4, Y1, Y3, Y2. The Y4 has been squeezed toward the y-axis in comparison to Y1, since it is happening twice as fast. The function Y2 is stretched twice as far from Y1 since it is happening half as fast.

The symmetry of stretching

In the previous example, you may feel that the whole function has just been shifted left since the values of the negative domain are hidden as they hug the x-axis. This is not the case, a stretch occurred away from the y-axis on both sides. To see this we graph the absolute value function and two related functions.

$$\text{(a) } f(x) = |x| \quad \text{(b) } g(x) = |2x| \quad \text{(c) } h(x) = |.5x|$$

Enter these functions as Y₁, Y₂, and Y₃. Set the viewing window to ZDecimal. We see that the graph of g is compressed on both sides, and the graph of h is stretched on both sides.

5.5 The Family of Quadratic Functions

The standard form of a quadratic equation is $y = ax^2 + bx + c$, where a, b, c are constants, $a \neq 0$. Letting $a = 1$ and $b = c = 0$, we have the simple quadratic equation $y = x^2$. It can be shown that any quadratic equation is just a transformation of this simple equation. The basic shape is called a parabola.

Transformation of the Basic Quadratic

[225] Let $f(x) = x^2$ and $g(x) = -2(x+1)^2 + 3$. Write g as a function of f. Graph f and g.

It is easy to see that $g(x) = -2f(x+1) + 3$. Enter that definition as Y₂ and press ZDecimal.

We see that g is a transformation of f. Also graph Y₃=-2X^2-4X+1 to see that it is the same as Y₂, thus, Y₃ is the standard form of Y₂.

Applications of Quadratics

Maximum/minimum

[229] For t in seconds, the height of a baseball in feet is given by the formula

$$y = f(t) = -16t^2 + 64t + 3.$$

Find the maximum height reached by the baseball.

Enter the formula in Y1 and graph with a viewing window of $0 \leq x \leq 4.7$ (Xscl=1) and $-20 \leq y \leq 100$ (Yscl=10). Use TRACE to find the maximum value as shown. By using the right half interval of the ZDecimal, the trace feature shows nice values of x. Good settings for the y values are found by trial-and-error or by using ZoomFit.

In some cases, the trace values may not land on the x value that gives the desired accuracy. In such cases, you can use maximum from the 2ndCALC menu. This process requires three prompts before it finds the maximum. You must specify:

Left Bound — a point to the left of the maximum

Right Bound — a point to the right of the maximum

Guess — a point near the maximum.

We see here the same answer. The maximum height is 67 feet. This maximum height is at the vertex of the parabola.

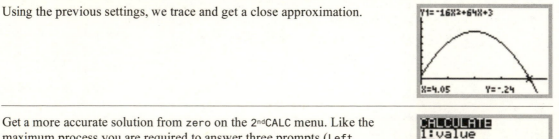

Tip: Finding a vertex (maximum/minimum) from a quadratic graph may show numbers that need some common-sense rounding. For example, it may find Y=2.99999999 when the exact answer is 3. In the above example, X=2.0000019 should be X=2.

The zeros of a quadratic

[218]   A second question from the baseball model is when does the ball hit the ground. This is found when $f(t) = -16t^2 + 64t + 3 = 0$, that is, when $f(t)$ has a zero.

Using the previous settings, we trace and get a close approximation.

Get a more accurate solution from zero on the 2ndCALC menu. Like the maximum process you are required to answer three prompts (Left Bound, Right Bound, Guess) before it finds the zero.

Notice that the y value is not exactly zero. But,

y=-1E-11=-0.00000000001 is pretty close.

The zeros of a quadratic function are found by using our program QF from Section 2.6.

Notes:

CHAPTER SIX

TRIGONOMETRIC FUNCTIONS

6.1 Introduction to Periodic Functions

Definition: A function is *periodic* if its values repeat at regular intervals. In functional notation this means that there is a c such that $f(t+c) = f(t)$ for all t in the domain of f. We see that the graph of the function repeats itself over successive intervals of length c. We start with an example of such a graph.

An Example of a Periodic Function

Show that the function $f(x) = x - \text{int}(x)$ is periodic. The function, int(x), is the integer part of x. For example, int(2.7128) = 2.

When entering the function definition for Y₁, the int(x) function needs to be pasted from the MATH NUM menu that is shown here, or it is found as int in the CATALOG.

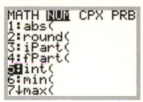

After entering the function, graph in a ZDecimal viewing window. For the TI-83, the screen shown here requires the MODE set to Dot instead of Connected. TRACE was turned on to display the function definition. The periodic behavior is evident.

Take special care with table values of periodic functions. Since the period is 1, if we set TblStart=0 and ΔTbl=1, then the table looks like the function is constant.

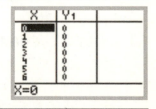

An Example of a Non-periodic Function

Enter Y₁=abs(int(X)) to see a function whose graph (in a ZDecimal window) is made up of repeated horizontal line segments of length 1. So it is clear that the only likely candidate for the period length is $c = 1$. However, it is not periodic since $f(x+1) \neq f(x)$.

6.2 The Sine and Cosine Functions

On a unit circle, each angle θ specifies a point, $P = (x, y)$. The sine function of the angle θ is the y-coordinate of the point P. The cosine function of the angle θ is the x-coordinate of the point P.

Degree Mode

Use the MODE key to change the angle mode to Degree. Note that only the TI-84 has the date/time line on the bottom of this screen.

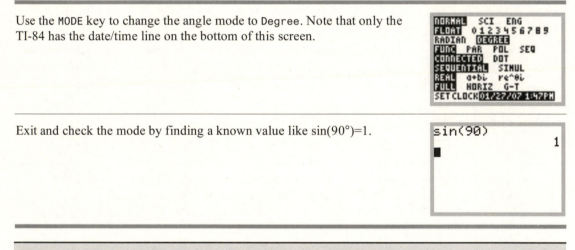

Exit and check the mode by finding a known value like sin(90°)=1.

Tip:	On many scientific calculators, the angle mode is shown on a status line across the bottom of the screen. There is no such indicator on the TI; you must use MODE to check.

The Ferris Wheel

[255]   A ferris wheel has a radius of 225 feet. Find your height above the ground as a function of the angle θ, measured from the 3 o'clock position. What is your height when the angle is 0°, 30°, 60°, ..., 360°?

We want the sequence of angle values in L1. There is an easy way to enter sequences into a list. From a clean home screen, press 2ndLIST and view the OPS menu as shown here. Select 5:seq(. The syntax is:

seq(*expression*, *variable*, *begin*, *end*, *increment*).

Complete the command as shown to store the list in L1. The same list could have been generated in an alternative way by using `seq(30X,X,0,12,1)`.

```
seq(X,X,0,360,30
)→L1
{0 30 60 90 120…
■
```

Next we want height values stored in L2. This screen shows the actual values in L2 which were generated by the formula $h = 225+225\sin(x)$ in L2.

L1	L2	L3	2
0	225	------	
30	337.5		
60	419.86		
90	450		
120	419.86		
150	337.5		
180	225		

`L2 =…5+225sin(X)`■

Check 2ndSTAT PLOT to set Plot1 as shown here. The plot is set On, it draws a scatter plot using L1 and L2, and the data points are square markers. If these are not the settings, then edit the Plot1 setup.

```
STAT PLOTS
1■Plot1…On
  ∟ L1   L2   □
2:Plot2…Off
  ∟ L1   L3   □
3:Plot3…Off
  ∟ L1   L2   □
4↓PlotsOff
```

Use ZoomStat to see a scatter plot. The axes appear as double lines because the Xscl and Yscl are so small that the tick marks form a second solid line on each axis.

Tip: The seq(command should not be confused with Seq as a function mode on the MODE screen.

6.3 Radians

In the previous section we stipulated that all the angles be in degrees. For calculus, radians are commonly used as a unit of angle measure. In this section we see how to convert back and forth between the two measures.

Converting Between Degrees and Radians

[259]   Convert 3 radians to degrees. Convert 3 degrees to radians.

To convert to degrees, check that the mode is set to Degree. Return to a fresh line on the home screen and enter 3.

```
3
```

Press 2ndANGLE and select 3:ʳ, which pastes the radian designation to the home screen.

You see that a 3 radian angle is about 172 degrees.

To convert degrees to radians, change to Radian mode. (Not shown.) Enter 3 and use 2ndANGLE and 1:°, which pastes the degree symbol to the home screen. You see that 3 degrees is about 0.05 radians.

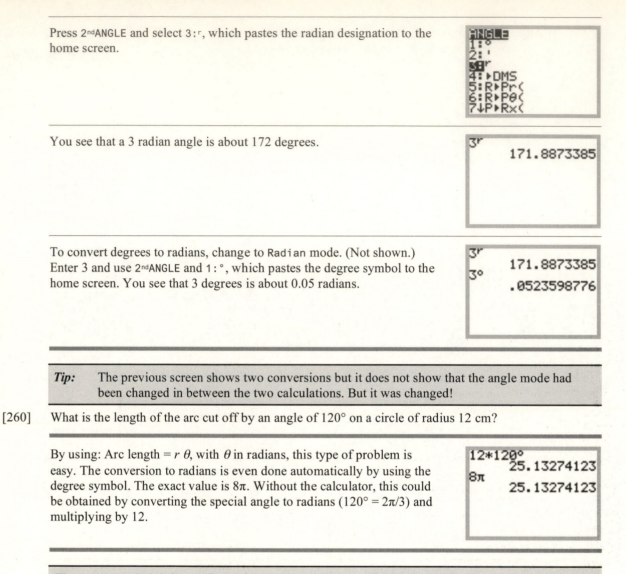

> *Tip:*   The previous screen shows two conversions but it does not show that the angle mode had been changed in between the two calculations. But it was changed!

[260]   What is the length of the arc cut off by an angle of 120° on a circle of radius 12 cm?

By using: Arc length = $r\,\theta$, with θ in radians, this type of problem is easy. The conversion to radians is even done automatically by using the degree symbol. The exact value is 8π. Without the calculator, this could be obtained by converting the special angle to radians ($120° = 2\pi/3$) and multiplying by 12.

> *Tip:*   As a rule it is best to leave the angle mode set to radians and write angles that are in degrees with the degree symbol designator (as shown in the previous screen).

6.4 Graphs of the Sine and Cosine

Exact Values for Sine and Cosine

[264]   In the last example of the previous section we saw that $8\pi \approx 25.13274123$. The calculator gave us an approximation instead of the exact form 8π. In exact terms, $\cos(30°) = \sqrt{3}/2$; however, the calculator shows $\cos(30°) = 0.8660254038$. In some cases, your instructor may want you to write an answer in exact form. More advanced calculators, like the TI-89 and TI-92, have the capability to leave answers such as $\cos(30°) = \sqrt{3}/2$ in exact form.

Graphing a Sine Function

[266]   Graph the ferris wheel function giving your height, $h = f(\theta) = 225 + 225\sin\theta$, in feet, above the ground as a function of the angle θ.

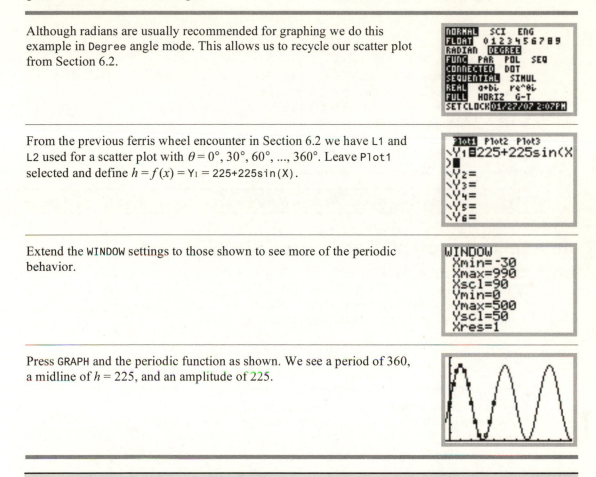

Although radians are usually recommended for graphing we do this example in Degree angle mode. This allows us to recycle our scatter plot from Section 6.2.

From the previous ferris wheel encounter in Section 6.2 we have L1 and L2 used for a scatter plot with $\theta = 0°$, $30°$, $60°$, ..., $360°$. Leave Plot1 selected and define $h = f(x) = $ Y1 = 225+225sin(X).

Extend the WINDOW settings to those shown to see more of the periodic behavior.

Press GRAPH and the periodic function as shown. We see a period of 360, a midline of $h = 225$, and an amplitude of 225.

Tip: For graphs of the pure trigonometric functions, such as $y = \sin(x)$ and $y = \cos(x)$, the ZTrig graph window is a good first setting. If MODE is set to Degree, the X settings are in degrees.

6.5 Sinusoidal Functions

Definition: A *sinusoidal* function is of the form

$$y = A\sin(B(t - h)) + k \text{ and } y = A\cos(B(t - h)) + k.$$

The general theory of transformations from Chapter 5 outlines the effect of the parameters on the basic sine or cosine function. We use graphing to compare the effect of a change in a parameter.

Horizontal Shifts

[272]   Compare the graph of $f(t) = \cos(3t)$ and $g(t) = \cos(3t - \pi/4)$.

Check the MODE to ensure a RADIAN setting. Enter the two functions, turn off Plot1 if on, and graph in a ZTrig window. TRACE shows that a maximum of the Y₁ graph occurs at $x = 0$.

Use the up arrow to change which function is being traced. Arrow right to see that the Y₂ graph has a maximum at $x \approx 0.26$. Thus, the second graph is the same as the first graph but shifted right by ≈ 0.26. Looking at the Y₂ definition, you might expect the shift to be $\pi/4$ (≈ 0.79). This example reminds us to factor when finding the horizontal shift. Here $g(t) = \cos(3(t - \pi/12))$ and $\pi/12 \approx 0.26$.

> *Tip:*   The ZTrig setting has an advantage in trace because it uses x-values that are fractions of π.

An Application of the Transformed Sine Function

[275]   In previous sections, we dabbled with the ferris wheel height problem by using a rotation angle. In reality, the height of the London ferris wheel is more naturally dependent on time. We now model height by using a sinusoidal function that assumes one revolution takes 30 minutes and that we board at the bottom. You can verify that the appropriate formula is

$$f(t) = 225\sin\left(\frac{\pi}{15}(t - 7.5)\right) + 225.$$

In Y₁ we have been cautious and inserted extra parentheses. Since we know that the midline is $y = 225$, we use Y₂ to graph it. The graph style of Y₂ is dotted.

Because it is an application, we have a good idea about the length of time (x-values) to show, say 120 minutes. In this example we also know the y-values because the wheel height is 450 feet. (As usual, we pad the top and bottom of the y settings.) We also set Xres=2. This is a feature that speeds up graphing by only plotting every other x-value.

The Xres (x-resolution) feature was used here not to speed up the graphing but to cause the midline to show as a dotted line. In dot style, a horizontal line looks like a connected solid line. The graph of Y₁ is what is important here. The midline graph is cosmetic. Using trace you can see your height as a function of time.

> *Tip:*   Xres is used to speed up graphing, but be careful about using a setting that is too high since it may distort a nonlinear graph. (Try Xres=5 in the previous case.) Reset Xres=1 after use. (ZTrig does not reset XRes.)

> *Tip:* Trigonometric graphs often show the pattern early in the graphing process. If you realize the graph is not what you want before it finishes, press ON to break.

6.6 Other Trigonometric Functions

In addition to the sine and cosine functions, there are four other common trigonometric functions.

Graphing the Tangent Function

[280] The tangent function is the quotient of the sine and cosine. This function is used enough that it has its own key, TAN. The TI-84 graphs $y = \tan(\theta)$ without difficulty. However TI-83 users see the following screen.

Enter the function in Y₁ and graph with ZTrig. This closely matches the textbook graph of the tangent function, but there is an important difference. The four vertical lines at $-3\pi/2$, $-\pi/4$, $\pi/4$, and $3\pi/4$ are dotted to indicate that they are vertical asymptotes. In this TI graph, the four vertical lines are connecting true discontinuities as explained in Section 2.3.

Change MODE from Connected to Dot and graph again. Use the right arrow to trace function values. The next point to the right, which is out of the window, is (1.44, 7.6). One more arrow to the right shows (1.57, -4.88). These two points were connected as a vertical line in the previous graph.

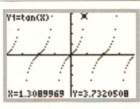

> *Tip:* You probably want to change the MODE back from Dot to Connected.

The Identity $\cos^2 x + \sin^2 x = 1$

[281] The identity $\cos^2 x + \sin^2 x = 1$ uses a notation that is not even possible to enter on a TI. Historically, this is shorthand, meaning $\cos^2 x = (\cos x)^2$. We must use parentheses for $\cos(x)$, and the squaring notation is entered after the closing parenthesis, not before it. See the examples below.

The expression $\cos^2 x$ is written in any of the following ways: cos(X)², (cos(X))², or cos(X)^2. The left side of the identity is written as Y₃.

We look at table values starting a 0 and increasing by 0.1. With a good eye for numbers, you can see the pattern that is displayed in Y₃, which is not yet shown.

Arrow over to Y₃. There it is! $\cos^2 x + \sin^2 x = 1$. This important identity is graphed in Section 7.3.

Reciprocal Functions

[282]   Each trigonometric function introduced so far has a reciprocal function,

$$\frac{1}{\cos(x)} = \text{secant}(x), \quad \frac{1}{\sin(x)} = \text{cosecant}(x), \text{ and } \frac{1}{\tan(x)} = \text{cotangent}(x),$$

but there are no secant, cosecant or cotangent keys on the TI. To use secant in a graph or calculation, you must write it as $1/\cos(x)$.

Enter the two functions. The style has been set to make the basic cosine graph thicker to distinguish the graphs. Use ZTrig.

The $1/\cos(x)$ is undefined when the $\cos(x) = 0$. This is precisely where there is a discontinuity. If you are using a TI-83 the connecting problem occurs for Y₂.

> *Tip:*   The COS⁻¹ key on your calculator is not the secant function. $\text{Cos}^{-1}(x) \neq \sec(x)$. This key is the inverse cosine and is introduced in the next section.

6.7 Inverse Trigonometric Functions

In Chapter 4 we encountered the logarithm as the inverse of the exponential function. The logarithm was used to solve exponential equations. In a similar fashion, we now introduce the inverse sine that is used to solve sine equations. The difference here is that trigonometric functions are periodic, so, for a given output, there are an infinite number of possible inputs that correspond to a given output value. To be a function, the correspondence must be unique, so we resolve this by restricting the interval of definition.

Solving Using the Inverse Cosine

[286]   Suppose we have a rabbit population modeled by the function $R = -5000\cos\left(\frac{\pi}{6}t\right) + 10000$. When does the population reach 12,000?

We start with an equation and solve for t

$$12000 = -5000\cos\left(\frac{\pi}{6}t\right) + 10000$$

$$-0.4 = \cos\left(\frac{\pi}{6}t\right)$$

$$t = \frac{6}{\pi}\cos^{-1}(-0.4).$$

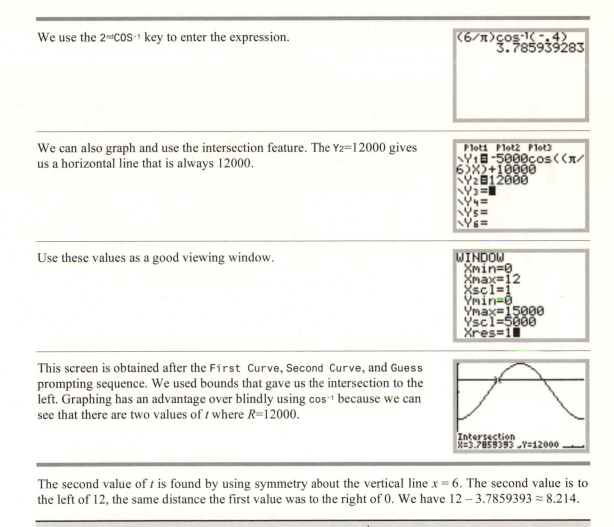

We use the 2ndCOS⁻¹ key to enter the expression.

```
(6/π)cos⁻¹(-.4)
          3.785939283
```

We can also graph and use the intersection feature. The Y₂=12000 gives us a horizontal line that is always 12000.

```
Plot1 Plot2 Plot3
\Y1=-5000cos((π/
6)X)+10000
\Y2=12000
\Y3=
\Y4=
\Y5=
\Y6=
```

Use these values as a good viewing window.

```
WINDOW
 Xmin=0
 Xmax=12
 Xscl=1
 Ymin=0
 Ymax=15000
 Yscl=5000
 Xres=1
```

This screen is obtained after the First Curve, Second Curve, and Guess prompting sequence. We used bounds that gave us the intersection to the left. Graphing has an advantage over blindly using cos⁻¹ because we can see that there are two values of t where $R=12000$.

```
Intersection
X=3.7859393  Y=12000
```

The second value of t is found by using symmetry about the vertical line $x = 6$. The second value is to the left of 12, the same distance the first value was to the right of 0. We have $12 - 3.7859393 \approx 8.214$.

Tip:	In mathematical writing the notation arcsin x and sin⁻¹x are interchangeable. The arcsin x notation is not permitted on the TI.

Tip:	Trigonometric functions can be used freely in Solver.

Graphs of the Inverse Trigonometric Functions

We now graph the three main inverse trigonometric functions. In each case a restriction to the domain of the original function has been made. We start with the $y = \cos^{-1}(x)$; note that this is not the same as the function $y = 1/\cos(x)$, which we call the secant.

Define Y₁=cos⁻¹(X) and graph in a ZTrig window to get a first look. We traced to explore the left endpoint on the graph. The graph is quite inadequate because the function is undefined over much of the screen.

The range of the cosine function is the domain of the inverse cosine, so we set $-1 \leq x \leq 1$. Also, the restricted domain of the cosine function is the range of the inverse function, so we set $0 \leq y \leq \pi$. The graph exactly fits the screen.

Using the same technique on $y = \sin^{-1}(x)$, we graph with the window set as $-1 \leq x \leq 1$ and $-\pi/2 \leq y \leq \pi/2$.

Since the range of the tangent function is all real numbers, the domain of the inverse tangent cannot be fully seen in a window.

We set $-10 \leq x \leq 10$ and use $-\pi/2 \leq y \leq \pi/2$.

CHAPTER SEVEN

TRIGONOMETRY

7.1 Laws of Sines and Cosines

In many triangle applications a right triangle is involved. In this section we look at solving triangle problems in general.

Law of Cosines with Solver

Solver is the handiest way to apply the Law of Cosines. However, if you have since used Solver with another equation, then you must rewrite the equation. We avoid this problem by storing the two laws in the bottom of the Y= menu. Since Y9 and Y0 are rarely used, they are good storage spaces for our formulas. The TI does not have upper and lower case variable names, so we name the sides A, B, and C and the corresponding angles X, Y, and Z. This correspondence must be remembered.

[309] A person leaves her home and walks 5 miles due east and then 3 miles northeast. How far away from home is she?

We enter the formulas in Y9 and Y0 as shown. After entry, the functions have been deselected so that they won't be graphed. (We want to leave them there forever, but we don't want them appearing in any graphs.) Note again that we have used the A-B-C for sides and X-Y-Z for corresponding angles when naming variables.

```
Plot1 Plot2 Plot3
\Y6=
\Y7=
\Y8=
\Y9=sin(X)/A-sin
(Y)/B
\Y0=C²-A²-B²+2A*
B*cos(Z)
```

Set the angle mode to Degree. (Not shown.) Use MATH Solver and arrow up to enter Y0, which has the Law of Cosines stored. (Y0 requires a paste from VARS Y-VARS.) Press ENTER.

```
EQUATION SOLVER
eqn:0=Y0█
```

Fill in the known variables. The angle is 135Arrow up to C, and make it a guess like 8. Press ALPHA SOLVE to calculate C.

The distance from home is about 7.43 miles, but the distance walked is 5+3 = 8 miles.

```
Y0=0
•C=7.4305587566…
 A=5
 B=3
 Z=135
 bound={-1ε99,1…
•left-rt=0
```

Tip: In the above example, you could avoid worrying about the angle mode by entering Z=135°. However, if you solve for Z, then the answer is given in the units corresponding to the angle mode setting.

Law of Sines

[311] Solve the ambiguous case of a triangle with sides $a = 8$, $b = 7$, and $\angle B = 40°$. If you draw a diagram, you see that $\angle A$ is either acute ($< 90°$) or obtuse ($> 90°$).

Before starting this example, it is noted that we use a different pasting technique than was used in the previous Solver example. Choose the way you like best. Start by clearing the Solver equation. Press 2ndRCL, and Rcl appears at the bottom of the screen as shown.
(The RCL is above the STO> key.)

```
EQUATION SOLVER
eqn:0=




Rcl ■
```

Now use VARS Y-VARS to paste Y₉ at the bottom of the screen. Press ENTER and the equation itself is pasted into Solver. The previous method only showed eqn:0=Y₉. The current method shows the actual formula and is thus safer.

```
EQUATION SOLVER
eqn:0=




Rcl Y₉
```

Now you explicitly see the Law of Sines, not the Y₉ variable. Remember our convention that the sides are A-B-C and the corresponding angles are X-Y-Z. (Sadly, this can't be avoided.)

```
EQUATION SOLVER
eqn:0=sin(X)/A-s
in(Y)/B



```

We solve for the acute angle first. Input A=8 for side a, Y=40 for angle B, and B=7 for side b. With angles in degrees, we find $\angle A = X \approx 47.3°$.

```
sin(X)/A-sin(…=0
■X=47.274644380…
 A=8
 Y=40
 B=7
 bound={-1E99,1…
■left-rt=0
```

Now reset X so that the initial guess is an obtuse angle. Notice how the black squares on the left disappear as an indication that this is no longer a calculated answer. In other words, don't believe the current X-A-Y-B values are a solution.

```
sin(X)/A-sin(…=0
 X=100■
 A=8
 Y=40
 B=7
 bound={-1E99,1…
 left-rt=0
```

The calculated answer shows $\angle A = X \approx 132.7°$.

```
sin(X)/A-sin(…=0
■X=132.72535561…
 A=8
 Y=40
 B=7
 bound={-1E99,1…
■left-rt=0
```

> **Tip:** If the two laws are left in Y₉ and Y₀ then they are brought back whenever needed. Just be sure that they are deselected (turned off) so that they do not affect graphing.

7.2 Trigonometric Identities

Identities are often used to simplify or to factor equations that are being solved. The TI is not of much use in this regard, but it does offer graphical techniques that can replace the need to solve algebraically. If answers need only be approximations, a graphical method is acceptable. In cases where the answer must be exact, the graphical approach is an excellent means of testing algebraically derived results.

[314]   Show the identity $\sin^2 t + \cos^2 t = 1$ graphically.

This is a fun surprise. Enter the left-hand side of the identity and graph in a ZDecimal window. The graph is clearly the constant function $y = 1$, which graphically verifies the identity. In Section 6.6 we also looked at this identity as a table.

[316]   Find all solutions to $\sin 2t = \sin t$, for $0 \le t \le 2\pi$.

Change to Radian angle mode. Enter Y₁=sin(2X) and Y₂=sin(X). Make the following WINDOW settings.

Press TRACE and you see the cursor in the middle of the screen. Luckily it is at one of the intersections. Applying common sense to the bottom line information ($x \approx \pi$ and $y \approx 0$) tells us that one of the solutions is $t = \pi$.

We can use the 2ndCALC intersect to find the next intersection to the right. Use a Guess about 5. These values are converted to exact values by noting that $y = -.8660254 = -\sqrt{3}/2$, which means $x = 5\pi/3$. Knowing symmetry, and aided by the graph showing five intersections, we list all the solutions:

$$t = 0, \ \pi/3, \ \pi, \ 5\pi/3, \ 2\pi.$$

[317]   Show the identity $\cos 2t = 1 - 2\sin^2 t$ graphically.

Enter the two functions as shown. The thick style is used for Y₂, so that we see visually that the second function is the same graph as the first.

We see a graphical verification that the equation is an identity. In this screen, the thick graph is seen as it is writing over the first graph.

Tip: Graphic verification is only a common-sense indicator. Formal proof requires algebra.

7.3 Sum and Difference Formulas for Sine and Cosine

Two of the most important trigonometric identities are

$$\sin(\theta + \phi) = \sin\theta\cos\phi + \sin\phi\cos\theta$$

$$\cos(\theta + \phi) = \cos\theta\cos\phi - \sin\phi\sin\theta.$$

By using these and the Pythagorean identity, numerous other identities can easily be derived. For example, let ϕ be negative and you have the difference formulas. Or let $\theta = \phi$ and you have the double-angle formulas. There are many more examples.

Your first inclination might be to hope $\sin(\theta + \phi) = \sin\theta + \sin\phi$. By trying $\theta = \pi$ and $\phi = 2\pi$, we see that it is true for these values. But remember an identity is true for all values. Try other values.

By trying $\theta = \pi/4$ and $\phi = \pi/3$, we see that $\sin(\theta + \phi) \neq \sin\theta + \sin\phi$. Next we verify the true formula for these values of θ and ϕ.

Here we see one example of values that make the equation true. But is it true for all values of θ and ϕ?

We can take a step closer to looking at all values of θ and ϕ by making θ a variable and letting ϕ be a constant. Our identity becomes $\sin(x + \pi/4) = \sin x\cos(\pi/4) + \sin(\pi/4)\cos x$. We use this to define the two sides of the equality as shown.

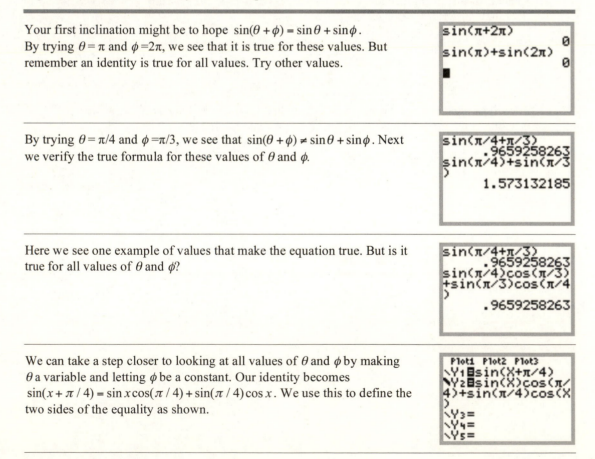

As strong graphical support for our identity, we are seeing the second thick graph trace over the first graph in a ZTrig window.

> *Tip:* For identity checking, the graphing style "bubble trace" is also a good choice. See the fifth style on page 19.

Rewriting sin(t) + cos(t)

[322] The previous graph showed that the Y2 function, which is essentially a sum of a sine and a cosine function, is equal to Y1, a single sine function. Using the sum identity with $\theta = x$ and $\phi = \pi/4$,

$$\sin(x + \pi/4) = \sin(x)\cos(\pi/4) + \sin(\pi/4)\cos(x) = (\sqrt{2}/2)\sin(x) + (\sqrt{2}/2)\cos(x).$$

Multiplying both sides by $\sqrt{2}$, we can write

$$\sqrt{2}\sin(x + \pi/4) = \sin(x) + \cos(x).$$

Enter the function Y1=sin(X)+cos(X) and graph in a ZTrig window. TRACE shows that the Y1 graph is not at a maximum when $x = 0$ and $y = 1$. The amplitude of Y1 must be greater than 1, and from $\sqrt{2}\sin(x + \pi/4)$, we know the amplitude is actually $\sqrt{2}$. The Y1 graph looks like a sine curve shifted left by some amount and from $\sqrt{2}\sin(x + \pi/4)$ we know the shift is $\pi/4$.

The general textbook result used to write the sum of a sine and a cosine as a single sine is:

$$A\sin(Bx + \phi) = a_1\sin(Bx) + a_2\cos(Bx)$$

where $A = \sqrt{a_1^2 + a_2^2}$ and $\tan\phi = \dfrac{a_2}{a_1}$.

7.4 Trigonometric Models

Dampened Oscillations

[331] A classic physics experiment is to collect data on the oscillation of a spring. A trigonometric function models the motion. However, this model may only fit in the short run since we know that the oscillations die out over time. We graph a model that both oscillates and dies out.

Our basic oscillation model is $y = \cos(2\pi x)$, which we dampens by using the exponential function $y = 5(1/2)^x$. The definitions of Y1 and Y2 form an upper and lower bound for the function model that is written as Y3.

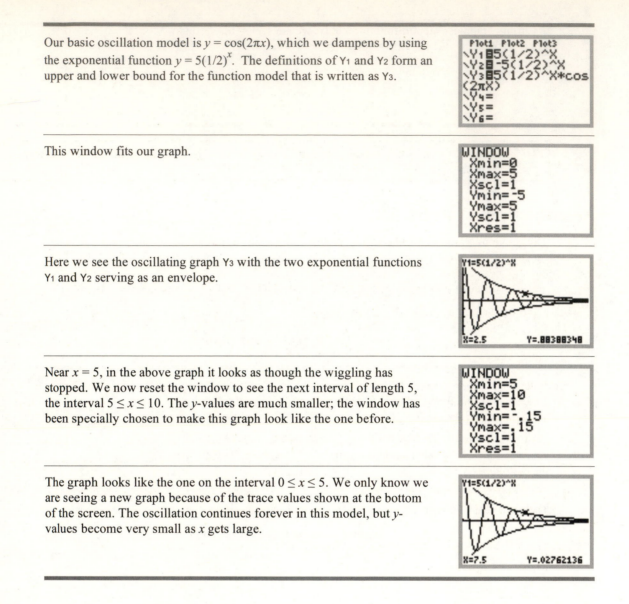

This window fits our graph.

Here we see the oscillating graph Y3 with the two exponential functions Y1 and Y2 serving as an envelope.

Near $x = 5$, in the above graph it looks as though the wiggling has stopped. We now reset the window to see the next interval of length 5, the interval $5 \le x \le 10$. The y-values are much smaller; the window has been specially chosen to make this graph look like the one before.

The graph looks like the one on the interval $0 \le x \le 5$. We only know we are seeing a new graph because of the trace values shown at the bottom of the screen. The oscillation continues forever in this model, but y-values become very small as x gets large.

7.5 Polar Coordinates

This section first introduces the procedure needed to convert between our standard Cartesian coordinate system and a polar coordinate system. (The text uses the term Cartesian, while the TI calls them rectangular coordinates.) Second, we look at the techniques for graphing a function that is defined in polar variables. We use only Radian angle mode in this section.

Converting Coordinates

The polar coordinate of a point P in the xy-plane is a pair r and θ, where r is the direct distance from P to the origin and θ is the angle measured counterclockwise from the positive x-axis to the line joining P and the origin.

[337]   Give Cartesian coordinates for the polar coordinate point $(7, \pi/3)$.
Give polar coordinates for the Cartesian coordinate point $(3, 4)$.

Conversion takes place on the home screen. We first press 2ndANGLE to see the conversion items; they are listed as 5:R▶Pr(through 8:P▶Ry(. They are also found in the CATALOG.

To get the x-coordinate, we paste the P▶Rx(on the home screen and complete the command with the polar coordinates as shown. To save some redundant entry for the y-coordinate, you can use 2ndENTRY and paste "P▶Ry(" over "P▶Rx(".

We continue and convert (3, 4) from rectangular coordinates to polar coordinates, obtaining the r-value and the θ-value, as shown.

Tip: In general, the TI uses the symbol ▶ to indicate translation/conversion expressions. We have seen ▶Frac, which converts a decimal to a fraction, and ▶Eff, which converts a nominal interest rate to an effective interest rate. Now we use P▶Rx, which converts a polar coordinate pair to the rectangular x-coordinate.

Tip: If you use the CATALOG, press Q and the up-arrow key to get to the P▶R listings more quickly than pressing P and using down arrow. Use S and the up-arrow key for the R▶P listings. But overall, the ANGLE menu seems the best choice.

Graphing Equations in Polar Coordinates

[338]   Next we see how to graph a spiral as a polar equation. The graphing modes (Func Par Pol Seq) have many similarities: the Y= menu keeps the function definitions, the viewing windows are set in WINDOW, TRACE is used to see function values, and TABLE shows lists of numerical values.

We use MODE to change the function mode from Func to Pol.

Press Y= and see that the function names start with r₁=. Polar equations are generally written as r in terms of θ (instead of y in terms of x). If the function mode is Pol, the variable key X,T,θ,n enters θ, not X. Or you can use ALPHA 3 to enter θ.

When in Pol mode, the window settings start with three new θ settings. These control the values plotted. Shown is the ZStandard window, and it says start at $\theta = 0$ and graph points until $\theta = 2\pi$. Each new point uses the previous θ plus the increment θstep. (The ZStandard value of θstep is $\pi/24$.) The actual window size is controlled, as before, with the X, Y settings.

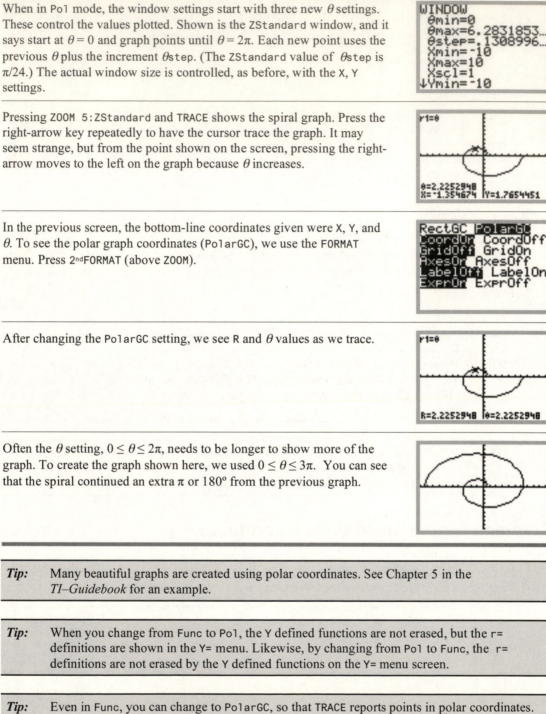

Pressing ZOOM 5:ZStandard and TRACE shows the spiral graph. Press the right-arrow key repeatedly to have the cursor trace the graph. It may seem strange, but from the point shown on the screen, pressing the right-arrow moves to the left on the graph because θ increases.

In the previous screen, the bottom-line coordinates given were X, Y, and θ. To see the polar graph coordinates (PolarGC), we use the FORMAT menu. Press 2ndFORMAT (above ZOOM).

After changing the PolarGC setting, we see R and θ values as we trace.

Often the θ setting, $0 \le \theta \le 2\pi$, needs to be longer to show more of the graph. To create the graph shown here, we used $0 \le \theta \le 3\pi$. You can see that the spiral continued an extra π or 180º from the previous graph.

Tip: Many beautiful graphs are created using polar coordinates. See Chapter 5 in the *TI–Guidebook* for an example.

Tip: When you change from Func to Pol, the Y defined functions are not erased, but the r= definitions are shown in the Y= menu. Likewise, by changing from Pol to Func, the r= definitions are not erased by the Y defined functions on the Y= menu screen.

Tip: Even in Func, you can change to PolarGC, so that TRACE reports points in polar coordinates.

7.6 Complex Numbers and Polar Coordinates

Definition: A *complex* number is defined as any number that can be written in the form

$$z = a + bi$$

where a and b are real numbers and $i = \sqrt{-1}$.

Tip: Do not confuse the *i*-vector from Chapter 10, with the number *i*.

Complex Numbers as Solutions to Equations

[342]   Solve $x^2 - 2x + 2 = 0$.

Recall in using the QF program, we got an error message in one case because the two calculations were not real numbers.

The program code is shown here as a reminder.

```
PROGRAM:QF
:Prompt A,B,C
:Disp (-B+√(B²-4
A*C))/(2A)
:Disp (-B-√(B²-4
A*C))/(2A)
:
:
```

Set the all the settings in the MODE menu to the default except change Real to a+b*i*.

```
NORMAL  SCI  ENG
FLOAT  0123456789
RADIAN  DEGREE
FUNC  PAR  POL  SEQ
CONNECTED  DOT
SEQUENTIAL  SIMUL
REAL  a+bi  re^θi
FULL  HORIZ  G-T
SET CLOCK 01/27/07 5:24PM
```

When the program is run and the values are input, the output is now two complex numbers instead of an error message.

```
prgmQF
A=?1
B=?-2
C=?2
                1+i
                1-i
               Done
```

Tip: If you use Solver, it does not return complex numbers as answers.

Algebra of Complex Numbers

[343]   Write $(2+7i)(4-6i) - i$ as a single complex number. Compute $(2+7i)/(4-6i)$.

We enter the complex numbers as given. (The *i* is a 2nd key above the period in the bottom row of the keypad.) The first calculation is clear. However, when we do the division, the answer is not exact and not completely visible on the screen. To correct this we press MATH 1, and the answer is translated to exact fractional form. (The Ans is automatically pasted when you press MATH 1 and does not need to be entered.)

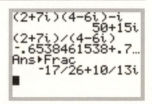

```
(2+7i)(4-6i)-i
             50+15i
(2+7i)/(4-6i)
-.6538461538+.7…
Ans▶Frac
        -17/26+10/13i
■
```

[345]   Convert 4^i to the form $a + bi$.

In this example, we see that complex numbers are even permitted as exponents. Because we expect decimal components that run off the screen we change MODE to show 3 rounded decimals.

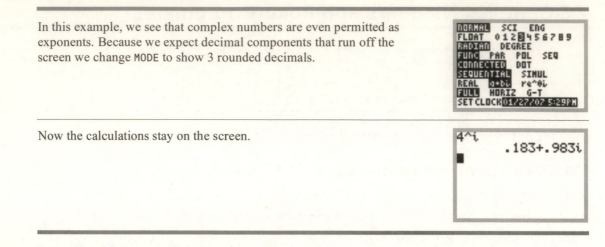

Now the calculations stay on the screen.

Polar Form of a Complex Number

Polar form is a compact way of expressing a complex number by its polar coordinates. In this subsection, for better screen output, we continue with the MODE set to 3-digit rounding. The angle mode is Radian.

[344]   Convert $z = -2i$ and $z = -2 + 3i$ to polar form.
Change the polar coordinates $r = 5$ and $\theta = 3\pi/4$ to complex form $a + bi$.

The MATH CPX menu has the two translation options 6: ▶Rect and 7: ▶Polar. These are pasted in expressions on next screen to make the desired conversions.

Enter complex numbers by using *i*. If you forget the *i* in either form, you get a DATA TYPE error. It is easy to forget in the polar form, so practice using e^(*i* together, as shown here.

| *Tip:* | When making frequent changes to MODE options, it is a good idea to check the settings before starting on any new problem, since the previous settings may no longer be desirable. You may also want to use 2ndFORMAT to check that RectGC is reset. |

CHAPTER EIGHT

COMPOSITION, INVERSES, AND COMBINATIONS OF FUNCTIONS

8.1 Composition of Functions

[356]   Let $p(x) = 2x + 1$ and $q(x) = x^2 - 3$. Suppose $u(x) = p(q(x))$ and $v(x) = q(p(x))$.
(a) Calculate $u(3)$ and $v(3)$.
(b) Find formulas for $u(x)$ and $v(x)$.

First check that all the MODE options are at their default settings. Now enter the functions p and q as Y1 and Y2, and then use these to enter u and v as Y3 and Y4. Since we only want values of u and v, we deselect Y1 and Y2.

Use 2nd TBLSET to start at 0 with ΔTbl=1. The table includes the values $u(3) = 13$ and $v(3) = 46$.

For part (b), the TI cannot give an algebraic formula, but you can check the graph of your algebraic derivation against the graph of Y3 and Y4.

Algebraically, we derive: $u(x) = 2x^2 - 5$ and $v(x) = 4x^2 + 4x - 2$. The Y3 graph is seen as $y = 2x^2$ shifted down by 5, which makes sense from the formula. The Y4 graph could be more easily checked if v was in vertex form, but we easily see that the y-intercept is –2.

Tip: You can shorten a composition definition by not showing the variable X. For example, Y3= Y1(Y2), but showing X is closer to function composition symbolism.

8.2 Inverse Functions

In Section 6.7, we introduced the inverse sine function $y = \sin^{-1}(x)$. The equation $\sin(x) = 0.8$ is solved by applying the inverse to both sides and using the TI to find $x = \sin^{-1}(0.8) \approx 0.927$. In this section, we look at inverse functions in general. If $y = f(x)$, then $x = f^{-1}(y)$. Inverse functions are functions in their own right and need not use the $^{-1}$ notation. For example, if $y = f(x) = 10^x$, then the inverse function is $y = \log(x)$. Not all functions have an inverse; those that do are called *invertible*.

An Invertible Function

[366]   Let $P(x) = 2^x$.
  (a) Show that P is invertible.
  (b) Graph P and P^{-1}.

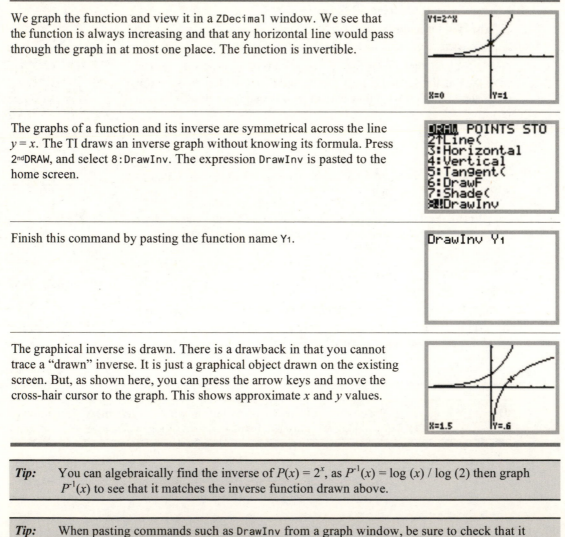

We graph the function and view it in a ZDecimal window. We see that the function is always increasing and that any horizontal line would pass through the graph in at most one place. The function is invertible.

The graphs of a function and its inverse are symmetrical across the line $y = x$. The TI draws an inverse graph without knowing its formula. Press 2ndDRAW, and select 8:DrawInv. The expression DrawInv is pasted to the home screen.

Finish this command by pasting the function name Y1.

The graphical inverse is drawn. There is a drawback in that you cannot trace a "drawn" inverse. It is just a graphical object drawn on the existing screen. But, as shown here, you can press the arrow keys and move the cross-hair cursor to the graph. This shows approximate x and y values.

Tip: You can algebraically find the inverse of $P(x) = 2^x$, as $P^{-1}(x) = \log(x) / \log(2)$ then graph $P^{-1}(x)$ to see that it matches the inverse function drawn above.

Tip: When pasting commands such as DrawInv from a graph window, be sure to check that it is pasted on a new line when you return to the home screen. If not, the pasted command continues on the command line and almost certainly causes a syntax error.

Checking Inverse Formulas

[368]   Check that $f(x) = \dfrac{x}{2x+1}$ and $f^{-1}(x) = \dfrac{x}{1-2x}$ are inverse functions of one another.

We start by entering the functions as Y1 and Y2. Set Y2 to thick style. Use ZDecimal. Shown here is Y1 being traced at the point $x = -0.5$. This point is not in the domain of Y1 since it has no y-value as shown. We strongly suspect that the two functions are inverses because they appear symmetric about the line $y = x$. (The symmetry line is not shown.)

A clever way to further test inverse formulas is to compose them. We know for inverse functions: $f(f^{-1}(x)) = x$ and $f^{-1}(f(x)) = x$. We write these as Y3 and Y4.

We see here that the graph of Y3 is the same as the graph of $y = x$. This is good visual proof of the inverse relationship, but again, only algebraic verification ensures it true for all x. This window setting is ZDecimal so that the $y = x$ line is at a 45° angle.

The inverse relationship holds for Y4 as well. It should be noted that Y1 is undefined at $x = -0.5$ and that Y4 is also undefined at this point. One pixel of the linear graph is seen missing there. In the previous screen, the missing point is seen at $x = 0.5$, where Y2 is undefined.

Tip: The missing points on the two linear graphs showed because the window setting ZDecimal caused x to be evaluated there. Other settings may not evaluate x and not show the undefined point.

Restricting Domains for Inverses

[370]   We have seen that the sine function must be restricted to the interval $-\pi/2 \le x \le \pi/2$ for it to have an inverse. We graph the full sine function and use DrawInv to see why the restriction is necessary.

We graph Y1=sin(X) (thick style) in a ZDecimal window and apply 2ndDRAW DrawInv. We see that the inverse is like a backward S and is certainly not a function. The inverse graph wobbles vertically between -1 and 1. TRACE is restricted to the Y1 graph.

Here we graph the arcsine function as Y2 and see that it does indeed have a restricted domain, $-1 \leq x \leq 1$, and a range $-\pi/2 \leq x \leq \pi/2$.

Using our composite checking technique, we see that, in the interval $-1 \leq x \leq 1$, the graph is on the line $y = x$.

Just for fun, instead of Y3=Y1(Y2(X)) you might check the other inverse relationship Y4=Y2(Y1(X)). What is happening here? This is not the line $y = x$! However, remember that the function Y1 is restricted to the interval $-\pi/2 \leq x \leq \pi/2$. On that interval, the graph coincides with the graph of $y = x$.

8.3 Combinations of Functions

We used combinations of functions, when we defined a sum and a quotient for trigonometric functions in Section 6.6. Now we look at combinations in general.

Graphs of Combinations

[373]   In Section 4.2 we introduced the population versus food model proposed by Malthus.

$$P = 2(1.04)^t \text{ and } N = 4 + 0.5t.$$

We now look at two ways to combine these functions to measure prosperity.

In Y3 the difference of the two functions is formed. When Y3 is zero, the food supply is the same as the population. After that point in time, a shortage occurs.

Graph the difference function with window $0 \leq x \leq 100$ and $-5 \leq y \leq 20$. We traced to get an approximation of the maximum difference, which occurs at about year 47. At this point there is enough food for almost 15 million additional people.

Now graph the quotient function Y4 with the window reset to $0 \le x \le 100$ and $-1 \le y \le 5$. We traced to get an approximation of the maximum quotient value, which occurs at about year 17. At this point there is enough food to feed over 3 times that population.

Thus, the two methods of measure, difference and quotient, give different answers for the time of maximum prosperity. However, to find when the food shortage begins, either graph yields the same answer, approximately 78 years.

Advanced Uses of Lists

The TI has the ability to name lists in addition to the standard lists, L1 to L6. We limit ourselves to looking at how to store a list by naming it and how to get it back as one of the L1 to L6 lists. Interested readers can find a fuller presentation of list features in Chapters 11 and 12 of the *TI-Guidebook*.

Naming a list

Suppose that L1 and L2 are the lists from Section 6.2, and we want to preserve this data, in order to use L1 and L2 for other data.

Use the STO> key to assign data from one list to a new list. The name for a list is up to five characters long, but must start with a letter. In this example we used the letter TR, since this was trigonometric data, and a number for the order. The list is displayed so that you can check what you have done.

Press 2ndLIST to see a list of currently stored list names. Your own screen depends on what lists are currently stored. This menu list is handy for pasting list names into commands. We continue back on our home screen in the next frame.

If a list named TR1 appears in the 2ndLIST NAMES menu, then it can be pasted on the home screen. Here TR1 is put into L3. To distinguish the list name TR1 from the name of a numeric variable the pasted list name is preceded by a small L. See the last command line.

In summary, we have moved a list of data from a standard named list and moved another list into this standard named list. You are allowed up to 20 list names to show in STAT Edit and they can be arranged in any order. We will use a simplified approach and view only the conventional six lists, L1 to L6. See Chapter 11 and 12 of the *TI–Guidebook* for details on list arrangements.

We separate the discussion of deleting lists, by TI model. The newer models are discussed first and the older model TI-83 (non-Plus) is discussed second. This is one of the rare occasions that there is a difference in screens between these models.

Notes on deleting lists and other stored items on a TI-84 and a TI-83 Plus

Press 2ndMEM to see a MEMORY menu of options. To clear all the lists at once, select 4:ClrAllLists. For this example we delete individual objects by using the option 2:Mem Mgmt/Del.

```
MEMORY
1:About
2:Mem Mgmt/Del…
3:Clear Entries
4:ClrAllLists
5:Archive
6:UnArchive
7↓Reset…
```

Next, we choose List as the kind of object to delete.

```
RAM FREE      22328
ARC FREE      36898
1:All…
2:Real…
3:Complex…
4:List…
5:Matrix…
6↓Y-Vars…
```

When the arrowhead ▶ is to the left of an item, it is ready to be deleted. Press DEL to delete. Be careful about your choice because once deleted the list is gone forever. If there is any doubt about what you are doing, use 2ndQUIT to get out.

```
RAM FREE      22328
ARC FREE      36898
    L1          129
    L2          129
▶   L3          129
    L4           12
    L5           12
    L6           12
```

A safer way to delete is by selecting one or more lists to be deleted. With the cursor on the name line, select the list name by pressing ENTER. An asterisk appears beside the name. The asterisk is a toggle it is turned on and off by ENTER.

```
RAM FREE      22451
ARC FREE      36736
    L1          129
    L2          129
  *L3          129
  *L4           12
▶ *L5           12
    L6           12
```

With the appropriate names selected by an asterisk, press DEL. You are asked to confirm before the deletion.

```
Are You Sure?
1:No
2:Yes
```

Tip: If you delete an L-list, it is also be deleted from the STAT Edit menu if it was showing there.

Tip   To restore the standard list order, press: STAT 5:SetUpEditor ENTER. This resets the six standard lists.

Notes on deleting lists and other stored items with a non-Plus TI-83

Press 2ndMEM to see a MEMORY menu of options. The menu is quite similar to the TI-84 and TI-83 Plus, but has fewer items. To clear all the lists at once, you select 4:ClrAllLists. As in the previous example we delete individual lists by using the option 2:Delete.

Like before we choose 4:List as the kind of object to delete.

When the arrowhead ▸ is to the left of an item, it is ready to be deleted. Press ENTER to delete. Be careful about your choice because once deleted the list is gone forever. There is no multiple selection process. If there is any doubt about what to do, use 2ndQUIT to get out.

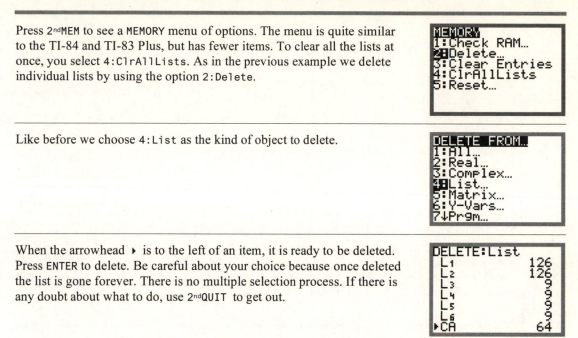

Combinations of Functions Defined by Tables

[377]   In this section, we look at two functions defined by data to form a quotient that represents a per capita crime rate.

Crime table

Year since 2000	0	1	2	3	4	5
Crime, City A	793	795	807	818	825	831
Crime, City B	448	500	525	566	593	652

Population table

Years since 2000	0	1	2	3	4	5
Population, City A	61000	62100	63220	64350	65510	66690
Population, City B	28000	28588	29188	29801	30427	31066

We enter the crime data in L1 and L2 and the population data in L3 and L4.

L1	L2	L3	1
793	448	61000	
795	500	62100	
807	525	63220	
818	566	64350	
825	593	65510	
831	652	66690	
-----	-----	-----	

L1(1)=793

We use L5 and L6 to calculate the ratios that represent the per capita crime rate. Move the cursor to highlight L5 and press ENTER. Enter the quotient formula as shown. When you pressed ENTER the list was calculated but the formula was lost.

L3	L4	L5	5
61000	28000	------	
62100	28588		
63220	29188		
64350	29801		
65510	30427		
66690	31066		
-----	-----		

L5 =L1/L3■

The previous technique could be repeated, but we show a different approach and store a formula in the header. In this screen you see that by enclosing the formula in quotes (") the formula itself is both applied and stored. A special symbol (■) appears in the header to alert you that it contains a formula. Move the cursor up to L6 to see the formula.

L4	L5	L6	6
28000	.013	.016	
28588	.0128	.01749	
29188	.01276	.01799	
29801	.01271	.01899	
30427	.01259	.01949	
31066	.01246	.02099	
-----	-----	-----	

L6 ="L2/L4"

We see that the per capita crime rate for City A starts at 1.3% and falls over the next five years, whereas the per capita crime rate for City B starts at 1.6% and rises over the same period.

Tip: To delete a list formula, enter a blank formula. For example, L6="".

CHAPTER NINE

POLYNOMIAL AND RATIONAL FUNCTIONS

9.1 Power Functions

Definition: A quantity y is proportional to a power of x if

$$y = kx^n,$$

where k and n are constant.

Directly Proportional

[388]   We compare three functions — the circumference of a circle, the area of a circle, and the volume of a sphere — each written in terms of its radius. These are all examples of a direct proportion.

$$C = 2\pi r \qquad A = \pi r^2 \qquad V = \frac{4}{3}\pi r^3$$

Graphs

Use Y= to show the formula screen. If you need to delete a previous formula, press the CLEAR key, with the cursor on the formula line. Here we show the three proportional functions that have been entered. Note that the = signs are highlighted to show these formulas are active.

```
Plot1 Plot2 Plot3
\Y1 ■2πX
\Y2 ■πX²
\Y3 ■(4/3)πX^3
\Y4=
\Y5=
\Y6=
\Y7=
```

We set graph style to distinguish between the three graphs. At the far left of each function definition is a graph style indicator. Arrow over and repeatedly press ENTER to change the style. We use regular, thick, and dot style as shown.

```
Plot1 Plot2 Plot3
\Y1 ■2πX
▶Y2 ■πX²
∙.Y3 ■(4/3)πX^3
\Y4=
\Y5=
\Y6=
\Y7=
```

Use WINDOW to set the viewing window values as shown.

```
WINDOW
 Xmin=0
 Xmax=6
 Xscl=1
 Ymin=0
 Ymax=100
 Yscl=10
 Xres=1
```

Press TRACE to see the graph and be ready to trace.

Inversely Proportional

[388]   The weight, *w,* of an object is inversely proportional to the square of the objects distance, *d*, from the earth's center:

$$w = \frac{k}{d^2} = kd^{-2}$$

Make a graph and table for the weight an object in terms of its distance from earth's center. On the surface of the earth, which is 3959 miles from the earth's center, it weighs 44 pounds.

Use the formula to find *k* from the values we know.

To make it more interesting we define the function with a fraction in Y₄ and with a negative exponent in Y₅

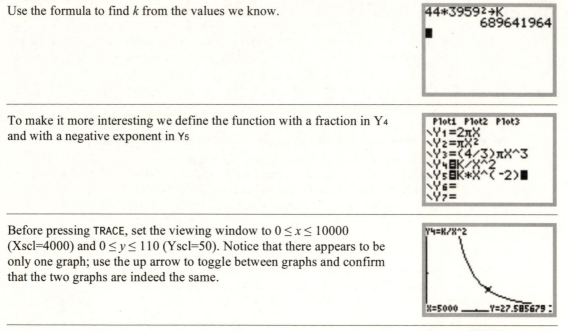

Before pressing TRACE, set the viewing window to $0 \le x \le 10000$ (Xscl=4000) and $0 \le y \le 110$ (Yscl=50). Notice that there appears to be only one graph; use the up arrow to toggle between graphs and confirm that the two graphs are indeed the same.

Check the table to further confirm that the definitions are the same. Press 2ndTBLSET and set TblStart=0 and ΔTbl=1000. Press 2ndTABLE to see the table shown. Notice that the two functions have the same output values. In addition you should note the ERROR entries on the top line. In this case, it tells you that division by 0 is an error in the formula.

Tip: Use common sense when interpreting a table with identical values or a graph that is the same in the viewing window. It may be a local phenomenon.

Definition: A *power* function is a function of the form

$$f(x) = kx^p, \text{ where } k \text{ and } p \text{ are constant.}$$

[390] You can explore various effects that k and p have on the family of power functions. For example, graph the four functions $y = x^{-1}, x^{-2}, x^{-3}, x^{-4}$, which you enter as Y1 to Y4, or as one function using set notation, Y1=X^{-1,-2,-3,-4}. Graph using a ZDecimal window to get a first view. The graph (not shown) shows a similarity between the even negative power functions and another similarity between the odd power functions.

To show the difference of behavior between odd and even powers, we investigate the graph of one odd representative, Y1=X^(-1), and one even representative, Y2=X^(-2). Technically you could enter these as Y1=X^-1, and Y2=X^-2 but the parentheses help make the expression clearer. We see from the graph that for negative x, the odd function is negative and the even function is positive. The window is ZDecimal.

If you are investigating the behavior of function values as they get close to a point (often zero), then first set Indpnt: Ask in the TBLSET menu. For example, after Ask is set, the table shown is constructed. The x-values approach zero from the positive side. No setting of ΔTbl could generate this table.

Note on fractional power graphs

Fractional powers, such as $x^{2/3}$ are acceptable, but they should be entered with parentheses to ensure proper meaning.

We enter the function Y1=X^(2/3) and graph with a ZDecimal window.

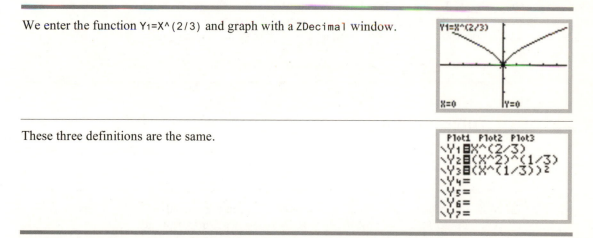

These three definitions are the same.

Tip: If you forget to use parentheses when entering a fractional power function, the graph is a power function of the numerator only. For example, the graph of Y1=X^2/3, is the parabola $y = (1/3)x^2$.

9.2 Polynomials

This section looks at polynomials. A polynomial is a sum of power functions whose exponents are nonnegative integers. The two issues addressed here are how to get a good graphing window and how to find the zeros of a polynomial.

Graphs of Polynomials

[399]   Show that the graph of $f(x) = x^3 + x^2$ resembles the power function $y = x^3$ on a large scale.

Enter Y₁=X^3+X^2 and use ZDecimal to get a first impression. (Not shown.) The behavior of the function near the origin is not shown very well. To get a better view we zoom in. Press ZOOM ZoomIn ENTER, then TRACE, to see the screen shown. Now we see the characteristic curve of a polynomial.

In the long run, a polynomial looks like the graph of $y = x^2$ or $y = x^3$, depending on its leading term. To see this behavior, we zoom out twice, once to get back to ZDecimal and once more to get the long-run view shown by this screen.

Press ZOOM ZoomOut ENTER ENTER, then TRACE.

The previous window showed us that this function looks like $y = x^3$ in the long run. To get a better feel for this, you would need to show more of the y-axis. If you enter Y₂=X^3 and reset the window to $-10 \leq x \leq 10$ and $-1000 \leq y \leq 1000$, then the graph of Y₁ is close to the graph of Y₂.

Zeros of Polynomials

[399]   Given the polynomial $q(x) = 3x^6 - 2x^5 + 4x^2 - 1$, find the intercepts. That is, where is $q(0)$ and for what x is $q(x) = 0$?

Enter the function and use a ZDecimal window. We can see immediately that the y-intercept is given by $q(0) = -1$. There are two x-intercepts at $x \approx \pm 0.5$. The x-intercepts are called zeros and are discussed in the next section.

9.3 Short-Run Behavior of Polynomials

We zoomed in and zoomed out on the graph of a function to see the short-run and long-run behavior of the polynomial. In many applications where polynomial functions are used as models, we need to find the zeros of the function. In the previous section, we saw that the zeros of a function are where the graph crosses the x-axis.

Graphs Showing Polynomial Zeros

[403] Investigate the short-run behavior of $u(x) = x^3 - x^2 - 6x$. Where does the graph cross the x-axis?

Enter Y1=X^3-X^2-6X and use ZDecimal to get a first impression. This is a poor view because we see three pieces of the graph. You may want to zoom out for a better view. However, we do see all three zeros in this window, so we estimate them from this window as $x = $ -2, 0, 3. You can evaluate Y1 to check that Y1(-2)=0, Y1(0)=0, and Y1(3)=0.

These zeros are quite obvious in this case. If the values are not obvious, you can find them by using 2ndCALC 2:zero.

The same procedure used for minimum, or maximum, is used at this point. You are prompted for

       Left Bound?, Right Bound?, Guess?.

With these suitably chosen, you can find the largest zero is at $x = 3$.

Multiple Zeros

[404] Describe the graphs of $y = (x+1)^2$, $y = (x+1)^3$, $y = (x+1)^4$, $y = (x+1)^5$, and so on.

This graph is seen as having a zero at $x = $ -1. The point of interest is that this differs from the zeros in the previous example because the graph does not continue through the x-axis. In nonmathematical terms, it just stops by the axis for a kiss.

In the next case, you can see that the graph *does* cross the x-axis. However, by close inspection, you can also see that the graph is quite flat near the zero.

Next we are back to a graph similar to the first graph. We can see a pattern that even powers look like a $\cup$ that touches the x-axis.

Finally, we conclude that odd powers pass through the x-axis but are quite flat at the zero. Since the function is factored as

$$y = (x+1)^5 = (x+1)(x+1)(x+1)(x+1)(x+1),$$

we see that $x = -1$ is a zero five times. It is a multiple zero.

9.4 Rational Functions

Definition: A *rational* function is one that can be written as a ratio of two polynomials. (We assume that the denominator is not the constant 0.) This definition means that even if a function is not in the form of a numerator over a denominator, it could still be considered rational if it could be written that way. For example,

$$f(x) = x + 3 - \frac{2}{x-2} \text{ is also written as } f(x) = \frac{x^2 + x - 8}{x - 2}.$$

The Long-Run Behavior

The long-run behavior of a rational function in fraction form is determined by the quotient of the highest power terms in the numerator and denominator.

[410]   Show that the long-run behavior of $f(x) = \dfrac{6x^4 + x^3 + 1}{-5x + 2x^2}$ is $y = 3x^2$.

Enter Y₁ with care. Use parentheses to enclose both the numerator and denominator, and be careful to use negation, not subtraction. Use thick style for Y₂=3X². Graph with ZDecimal and TRACE. We see a poor picture for the long-run. We need to expand the window. *Note:* The Y₁ equation is too long to be completely shown at the top of the screen.

In this window we see how the long-run values at either side of the graph are similar. We ignore the spike in Y₁ because it is short-run behavior. The settings for this window were $-20 \le x \le 20$ and $-200 \le y \le 1000$.

Tip: In defining a rational function, be sure to include parentheses. Use the form Y =(…)/(…) even when it is not necessary.

[411]   Show that the long-run behavior of $r(x) = \dfrac{x+3}{x+2}$ is $y = 1$. (This could also be stated as: 1 is a horizontal asymptote.)

Enter the function and use ZStandard to see that the function values approach 1 as $x \to \pm\infty$. We again ignore the spike in Y3 at x=-2 because it is short-run behavior. The TI-83 may connect spike.

Tip: For Hollywood-style functions that have all their zeros and asymptotes close to the origin, the ZStandard window often allows a good view of both short- and long-run behavior.

[413]   Show that the long-run behavior of $r(x) = \dfrac{1}{x^2 + 3}$ is asymptotic to $y = 0$. We also could have said that the behavior of r is like that of $y = 1/x^2$ (which has a horizontal asymptote at zero).

Enter the function and use ZStandard. A better window is needed for the short-run behavior, but it does appear that the function values approach 0 as $x \to \pm\infty$. (You may need to trace left and right to convince yourself.)

Changing the y settings to $-0.5 \le y \le 0.5$ gives a better view of the function values as $x \to \pm\infty$. They approach 0 as a horizontal asymptote. In contrast to the other examples of this section, there is no vertical asymptote. Only rational functions with a denominator that is zero for some x can possibly have a vertical asymptote.

9.5 Short-Run Behavior of Rational Functions

The rational functions are often the hardest graphs to get a good viewing window because you want to see both the long- and short-run behavior. By finding the zeros of the numerator, you have all the possible zeros of the rational function. By finding all the zeros of the denominator, you have all the possible vertical asymptotes. Combine this with the method of finding long-run behavior from the last section and you should have a good idea of how the graph of a given rational function should look.

The Graph of a Rational Function

Describe the graph of rational function

$$g(x) = \frac{3x+2}{(x-1)(x+2)}.$$

Enter the function and view in a ZStandard window. We see the overall pattern we expected: one zero at -2/3, two vertical asymptotes at $x = -2$ and $x = 1$, and a horizontal asymptote at $y = 0$. For TI-83 users, lines connect discontinuities and appear like asymptotes.

A TI-83 purist would be upset by the connection of the graph across the two vertical asymptotes. This can be avoided by changing Connected to Dot in the MODE menu. However, as you can see, the graph is less suggestive of the continuity that does exist near the two vertical asymptotes.

For the TI-83, another means of avoiding the dishonest connection but staying in the Connected mode is to be sure that the function is evaluated at undefined points. In this ZDecimal window (in Connected mode), the connections did not take place because both $x = -2$ and $x = 1$ were evaluated as undefined points. Graph connections are not made across points evaluated as undefined.

By now you should be sophisticated enough to know that the connection lines across undefined values of a graph should not really be there. It is easier to see them for what they are and to ignore them. If you are a purist, change modes or finding special viewing windows to avoid connecting lines.

[418]   Describe the graph of rational function

$$r(x) = \frac{x^2 + 2x - 3}{x^2}.$$

Enter the function and view in a ZStandard window. We see the overall pattern: one zero at -3, the other at 1. The function is undefined at $x = 0$ and has a vertical asymptotes there. There is a horizontal asymptote at $y = 1$.

View the same function again in a ZDecimal window. The horizontal asymptote appears to be greater than 1 in this window. We must exercise care when finding asymptotes from graphs.

We added the horizontal line Y₂=1 and use a window of $-20 \le x \le 20$ and $-1.5 \le y \le 1.5$. Now we see the behavior of the graph more clearly. Near the origin it crosses the asymptote and then slowly gets close to $y = 1$ from above. Where as for negative x it approaches 1 from below.

When Numerator and Denominator Have the Same Zeros

[419]   Describe the graph of rational function

$$h(x) = \frac{x^2 + x - 2}{(x-1)}.$$

We first graph this function in a ZStandard window. The zeros of the numerator are $x = -2$ and $x = 1$, but we see graphically that there is no intercept at $x = 1$, where we expect it. But note the denominator has a zero at $x = 1$, and there is no vertical asymptote. Thus, there is no zero at $x = 1$ because it is undefined.

Using a ZDecimal window, we see that there is a pixel missing in the graph at $x = 1$. Trace to confirm this. In the ZStandard window we could not trace to the exact point $x = 1$, and so there was no missing pixel to warn us about the undefined point.

Tip: When copying a graph such as the one above, you should draw an open circle at the undefined point. This was previously discussed at the end of Section 2.3.

9.6 Comparing Power, Exponential and Log Functions

In this section, we graph and compare three different families of functions. The definition and basic graphing are straightforward. However, you may want to review some techniques that allow you to find good viewing windows. In some cases with multiple functions, there is a good viewing window for one but not all of the functions, and in the end there may not be one window that meets the needs of all the functions in a particular graph.

[424]   Compare $f(x) = x^4$ and $g(x) = 2^x$.

We graph the functions in a ZDecimal window. Looking at the screen we see that the exponential function has larger values near the origin. We know $g(x)$ has a zero horizontal asymptote as $x \to -\infty$, so $f(x)$ obviously dominates for negative x. It is unclear which function might dominate as $x \to \infty$.

To find the long-run behavior, we look at table values to give us a clue. At $x = 10$, we see that Y2 lags Y1, but by $x = 20$, Y2 is near a million and seems to dominate Y1 from then on.

(Recall that in E–notation, 1.05E6 = 1,050,000.)

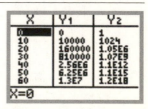

To be sure about the end behavior for positive x, we continue for even larger x-values. Yes, we see that Y_2 continues to dominate, but then we encounter ERROR. This is because the Y_2 function values exceeded the limit of the calculator. (No exponent can exceed two digits.)

To set a nice window, we use the million value obtained from our table with $x = 20$. The Ymin is set to allow room at the bottom of the screen for trace values.

After graphing, it may not be clear which function is which, so we use TRACE and arrow over to find that the power function is dominated by the exponential function. This is a general result: Any exponential function eventually dominates any power function.

This final screen, which is essentially the one above, could have been more quickly by setting $0 \leq x \leq 20$ and using Zoomfit. It avoids looking at table values to decide good y window settings.

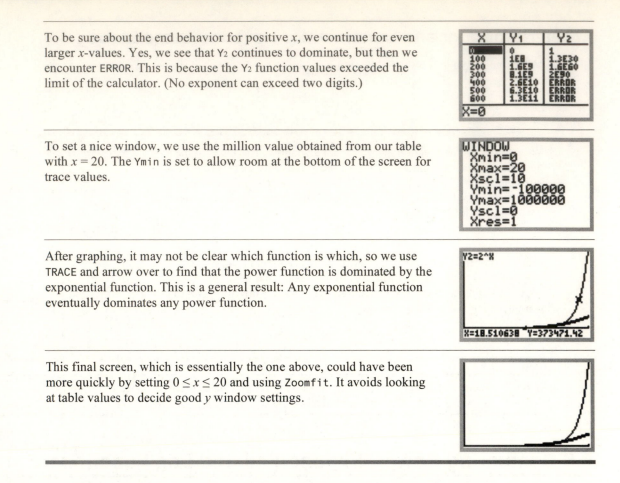

9.7 Fitting Exponentials and Polynomials to Data

In Section 1.6 we saw how to fit a linear regression equation to data. In this section we show how to fit either an exponential regression equation or a power regression equation to data.

The Spread of AIDS

[428]   The table below gives the total number of deaths in the United States from AIDS since 1981. We use $t = 1$ for 1981.

t	1	2	3	4	5	6	7	8
N	159	622	2130	5635	12607	24717	41129	62248

t	9	10	11	12	13	14	15	16
N	90039	121577	158193	199287	243923	292586	340957	375904

One approach to fitting a curve to this data is to use the log function to *linearize* the data. This yields a linear regression equation for the converted data that is used to algebraically derive the exponential or power regression formula. This approach is good for conceptual understanding, and your instructor may want you to do the work that way. However, the TI can derive both exponential and power regression formulas in a simple sequence of steps.

Exponential Regression

Enter the data using STAT Edit. The data entry in L1 is not time-consuming but using seq(is faster. See Section 6.2.

Turn off any active Y= functions. Use 2ndSTATPLOT to turn on Plot1 and ZoomStat to see a data plot. It is clear that a linear regression equation does not seem appropriate.

On a fresh home screen, use STAT CALC and arrow down to select ExpReg (Exponential Regression) to begin the command.

After pasting ExpReg, specify the two list names and a Y= equation for the regression formula to be pasted into.

The parameters a and b are calculated for the equation y=a*b^x. The r and r^2 are measures of correlation as previously described.

Optional step: We check that Y1 has been properly pasted and that both Y1 and Plot1 are selected to be graphed.

This graph shows the data points and the exponential regression curve. The approximation starts out well, but then is a little low in the middle and appears to be too high for larger values. We now try to do better with a power regression equation.

Power Regression

Now select PwrReg from the STAT CALC menu and create the command: PwrReg L1,L2,Y1.

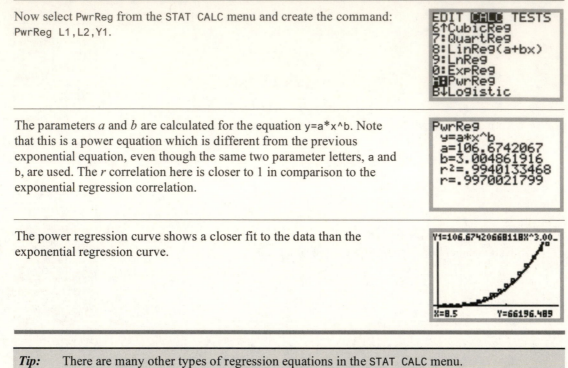

The parameters *a* and *b* are calculated for the equation y=a*x^b. Note that this is a power equation which is different from the previous exponential equation, even though the same two parameter letters, a and b, are used. The *r* correlation here is closer to 1 in comparison to the exponential regression correlation.

The power regression curve shows a closer fit to the data than the exponential regression curve.

Tip: There are many other types of regression equations in the STAT CALC menu. For a complete description, see Chapter 12 in the *TI–Guidebook*.

CHAPTER TEN

VECTORS AND MATRICES

10.1 Vectors

The TI does not support vector definitions or vector operations, but by thinking of a vector as a list of components, we can use the list features of the TI to work with vectors. Both the TI-89 and the Voyager 200 have a vector data type and allow some vector operations.

Addition of Vectors

[450] A vector can represent strength and direction of a gravitational force. A spacecraft, thought of as situated at the origin, may experience two vectors of force. One force is eight units along the *x*-axis, and the other is three units at an angle of -70° with the *x*-axis. Find the magnitude and direction of the sum of the two force vectors.

Moving the tail of the second vector to the head of the first vector gives a triangle with adjacent sides of 3 and 8 and an included angle of 110° (110° = 180° – 70°). We use the Law of Cosines and Law of Sines from Section 7.1 to find our results.

Recall that we have the Law of Cosines stored in Y0, so we use Solver to find the length of the third side, which in this case is the magnitude. The steps, which are not shown, are:

1. Set mode to Degree
2. Paste Y0 as the equation
3. Use down-arrow to enter the values of A, B, and Z
4. Arrow up to C
5. Press ALPHA SOLVE.

The magnitude of the sum is C.

To find the angle of declination for the sum vector, change the equation to the Law of Sines (Y9) and down-arrow to enter the values of X, A, and B. You see that A has full accuracy, which certainly was not entered by hand. By entering A=C, the value was pasted from the previous result. As usual, the angle value, Y, is found by setting it to zero, as an estimate, and pressing ALPHA SOLVE.

The above approach is overly complicated and is simplified in Section 10.3.

10.2 Components of a Vector

There are several ways a vector may be designated. A so-called free vector is given by a pair of rectangular coordinates for its head and another pair of coordinates for its tail. The components of a vector are the coordinate pair given as if the vector had moved its tail to the origin. The polar coordinates of a point also describe a vector whose tail is at the origin.

To facilitate finding components and working with vector operations, we create a program to use. To enter the program requires extensive pasting and entering of ALPHA characters. It is easier to use *TI-Connect™* and enter the program from the keyboard of a computer. It is even easier to have a friend or the instructor let you copy the program from their computer or from their TI by using the 2ndLINK command. (See the Appendix for details.) The creation and naming of a program were introduced in Section 2.6.

After a program has been named and entry begun, you can exit at any time with 2ndQUIT. To check the program code or continue editing, press PRGM EDIT and select the desired number.

```
EXEC EDIT NEW
1:FOCALE
2:QF
3:VECTORS
```

This screen contains the first few lines of the program so that you can check your work for proper entry. Long lines are wrapped for a full view. New lines start with a colon. Exit with 2ndQUIT.

```
PROGRAM:VECTORS
:Menu("VECTORS",
"POSITION",A,"R,
θ",B,"I,J",C,"EX
IT",E
:Lbl A
:Input "X1? ",X
:Input "Y1? ",Y
```

Vector Program

```
Menu("VECTORS","POSITION",A,"R,θ",B,"I,J",C,"EXIT",E
Lbl A
Input "X1? ",X
Input "Y1? ",Y
Input "X2? ",Z
Input "Y2? ",W
Z-X→I
W-Y→J
R ▶ Pr(I,J) →R
R ▶ Pθ(I,J) →θ
Goto D
Lbl B
Input "R? ",R
Input "θ? ",θ
P ▶ Rx(R,θ) →I
P ▶ Ry(R,θ) →J
Goto D
Lbl C
Input "I? ",I
Input "J? ",J
R ▶ Pr(I,J) →R
R ▶ Pθ(I,J) →θ
Lbl D
ClrHome
Disp "R,θ",R,θ
Disp "I,J",I,J
Lbl E
Stop
```

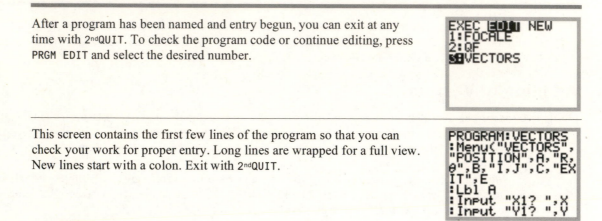

Help to find commands:	
"	ALPHA +
θ	ALPHA 3
ClrHome	PRGM I/O 8
Disp	PRGM I/O 3
Goto	PRGM CTL 0
Input	PRGM I/O 1
Lbl	PRGM CTL 9
Menu(	PRGM CTL C
P ▶ Rx	2ndANGLE 7
P ▶ Ry	2ndANGLE 8
R ▶ Pr	2ndANGLE 5
R ▶ Pθ	2ndANGLE 6
Space	ALPHA 0
Stop	PRGM CTL F

The basic idea of this program is to show a menu of three ways to represent a vector. We enter known coordinates in one of the three representations and the vector's coordinates are shown in two forms. Input choices are by POSITION, meaning a free vector; by R, θ, meaning polar coordinates; or by I,J, meaning rectangular components. The output screen shows both the R,θ and the I,J coordinate representation.

> *Tip:*   If you use the *TI-Connect*™ to enter a program from your computer, be careful about adding spaces at the end of command lines. After the program is transferred to your calculator, extra (unseen) spaces cause syntax errors.

Changing from Polar to Rectangular Coordinates

[456]   A ship travels 200 miles supposedly going due east, but in fact it has traveled 17.4° to the north of east. How far north is the ship from its intended course?

     We enter the vector as R and θ and use the program conversion to find J, the north component.

To start the program, press PRGM EXEC and select the program.

The program name is pasted to the home screen. Press ENTER.

The program menu is shown; select 2:R, θ.

Enter the R and θ as shown. If you are not in Degree mode, then you should enter θ=17.4°. Even when you are entering data within a program, you can use 2ndANGLE to paste the degree symbol.

The answer is J ≈ 59.8. The values shown for R, θ, I, and J are stored in memory, so if needed, they are available for further calculations.

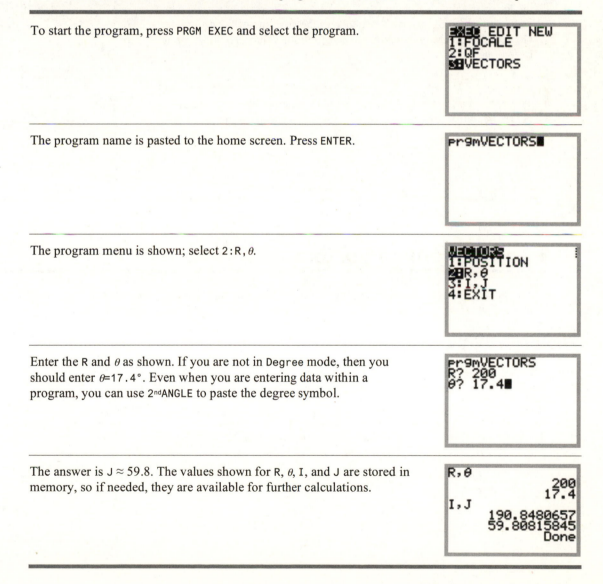

> *Tip:*   After you see the program output on the home screen, you can press ENTER and the program starts again. This is very handy for dealing with several vectors.

Finding Vector Components

[459]   Find components of a vector whose tail is (-3,1) and whose head is (2,4).

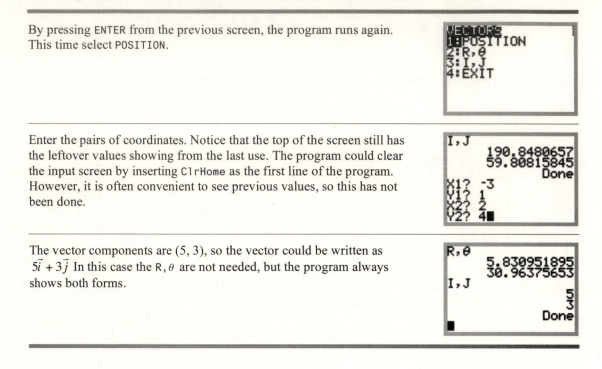

By pressing ENTER from the previous screen, the program runs again. This time select POSITION.

Enter the pairs of coordinates. Notice that the top of the screen still has the leftover values showing from the last use. The program could clear the input screen by inserting ClrHome as the first line of the program. However, it is often convenient to see previous values, so this has not been done.

The vector components are (5, 3), so the vector could be written as $5\vec{i} + 3\vec{j}$ In this case the R, θ are not needed, but the program always shows both forms.

10.3 Applications of Vectors

So far the vectors have been only two-dimensional and handled by using a variable for each dimension, as seen by our use of I and J. For higher dimensional vectors, we use lists.

Vector Arithmetic

[463]   Suppose $\vec{R} = (domestic, international, videotape)$ is a revenue vector, in millions of dollars, for a movie. The movies *Shrek*, *Pearl Harbor*, and *The Mummy Returns* had vectors:

$$\vec{s} = (268, 198, 436), \ \vec{t} = (199, 252, 301), \text{ and } \vec{u} = (202, 227, 192).$$

Find the total revenue vector for all three of these movies.

Enter the data for each vector in a list.

Arrow to the heading for L4 and press ENTER. Enter the sum of the three lists as shown on the bottom line.

L4 is the total revenue vector.

[465] In computer graphics, a pixel has an (x, y) location and is often thought of as being moved by a given distance in a certain direction. This movement is described by a vector (r, θ). For example, suppose the point $(3, 5)$ is moved three units in the 70° direction where the angle is measured counter-clockwise from the x–axis.

Start the VECTOR program and select 2:R,θ from the menu.

Enter the values for R=3 and θ=70 as shown on the screen.

Pressing ENTER gives the I,J components of the movement. These components are added to the original pixel point position to find the new position.

Since the original pixel position in I,J form. Start the program again by pressing ENTER and select 3:I,J.

Enter the value of I as ALPHA I + 3.
Enter the value of J as ALPHA J + 5.

This reuses the full accuracy of the previous I and J as shown on the
next screen.

```
                         3
                        70
I,J
          1.02606043
          2.819077862
               Done
I? I+3
J? J+5█
```

The final output screen shows the new position of the point as I,J.

```
R,θ
          8.794722349
          62.75596142
I,J
          4.02606043
          7.819077862
               Done
```

[450]   In a similar, way we can redo the rocket vector sum example from Section 10.1. The spacecraft,
situated at the origin, experiences two vectors of forces. One force is eight units along the *x*-axis, and
the other, is three units at an angle of -70° with the *x*-axis. We want to find the magnitude and direction
of the sum of the two force vectors. The new approach is a repeat the previous sequence of steps and
adds the components of the two vectors to find the resultant vector. We avoid having to think about
moving vectors, forming triangles, and using both the Law of Sines and the Law of Cosines. This
approach is rocket science made simple.

Select the 2:R,θ on the program menu screen. Enter R=3 and θ=70 to
display the screen shown. Think of declination as positive to avoid
negative signs.

```
R,θ
                         3
                        70
I,J
          1.02606043
          2.819077862
               Done
█
```

Press ENTER to start the VECTOR program again. Select 3:I,J. We add the
horizontal vector (I=8 and J=0) to the previous vector, whose
coordinates are still stored in I and J.

Enter the value of I as ALPHA I + 8.
Enter the value of J as ALPHA J.

```
                         3
                        70
I,J
          1.02606043
          2.819077862
               Done
I? I+8
J? J█
```

We read the sum using the R,θ vector. This corresponds to the result
found in Section 10.1. We have a force of R ≈ 9.456 acting at an angle of
declination θ ≈ 17.345.

```
R,θ
          9.456054509
          17.34496778
I,J
          9.02606043
          2.819077862
               Done
```

Tip: After you see the program output on the home screen, you can press ENTER and the program
menu shows again. Press 4:EXIT to return to normal use of the home sceen.

10.4 The Dot Product

Definition: If $\vec{u} = (u_1, u_2, \cdots, u_n)$ and $\vec{v} = (v_1, v_2, \cdots, v_n)$ are *n*-dimensional vectors, then the *dot product* is defined as

$$\vec{u} \cdot \vec{v} = u_1 v_1 + u_2 v_2 + \cdots + u_n v_n.$$

[468] A car dealer sells five models of a car. The number of cars sold in a week is given by the vector $\vec{C} = (22, 14, 8, 12, 19)$, and the price of each model is given by the vector $\vec{P} = (14, 18, 35, 42, 27)$. Find this week's revenue.

On the home screen, enter the vectors as lists. (You could also use STAT Edit.)	`{22,14,8,12,19}→L₁`
	`{22 14 8 12 19}`
	`{14,18,35,42,27}`
	`→L₂`
	`{14 18 35 42 27}`
	`▪`

From the 2ndLIST MATH menu select sum(.	`NAMES OPS MATH`
	`1:min(`
	`2:max(`
	`3:mean(`
	`4:median(`
	`5▪sum(`
	`6:prod(`
	`7↓stdDev(`

Complete the sum command as shown with a multiplication of the two lists. Multiplication of lists is term by term, so by summing the multiplication list you have the dot product.	`L₁`
	`{22 14 8 12 19}`
	`{14,18,35,42,27}`
	`→L₂`
	`{14 18 35 42 27}`
	`sum(L₁*L₂)`
	`            1857`

[470] Find $\vec{v} \cdot \vec{w}$ where $\vec{v} = 3\vec{i} + 4\vec{j}$ and $\vec{w} = 2\vec{i} + 5\vec{j}$. Find $\| \vec{v} \| = \sqrt{\vec{v} \cdot \vec{v}}$ and $\| \vec{w} \| = \sqrt{\vec{w} \cdot \vec{w}}$.

We find the dot product using lists. However, it is hardly worth putting two-dimensional vectors into lists. In this simplistic case, you could enter the calculation directly from the dot product definition to the home screen and get, 3*2+4*5 = 26.	`{3,4}→L₁`
	`           {3 4}`
	`{2,5}→L₂`
	`           {2 5}`
	`sum(L₁*L₂)`
	`              26`

We find the norm (or length) of a vector by using the formula as shown on the screen; this applies no matter what the dimension of the vector.	`√(sum(L₁*L₁))`
	`            5`
	`√(sum(L₂*L₂))`
	`   5.385164807`

In the two-dimensional vector case, the norm is just the R of the polar coordinate. To show this, we use the VECTORS program; select 3:I,J. Enter I=2 and J=5 for the second vector to see that R=5.385... is the norm of the vector.

Tip: Values shown on the VECTORS program screen are stored in variable names and remain until the next calculation of the program or changes are made to that variable. A value from a home screen calculation (in Ans) remains only until the next calculation.

10.5 Matrices

Tip: The matrix menu key on the TI-83 (non-Plus) has been replace by the APPS key on the TI-83 Plus and the TI-84. The TI-83 (non-Plus) users should translate references to 2ndMATRIX to be the MATRX key. There is no I in the label MATRX on the TI-83 (non-Plus).

Matrix Arithmetic

A matrix is a rectangular array of numbers. It can, like vectors, be multiplied by a scalar. Matrices of the same dimensions can be added and subtracted.

Addition and scalar multiplication

[481]   Evaluate the expressions (a) $5A$ (b) $-2B$ (c) $A+B$ (d) $B-3A$, given that:

$$A = \begin{pmatrix} 3 & 7 \\ 2 & -1 \end{pmatrix} \text{ and } B = \begin{pmatrix} 1 & -5 \\ 0 & 8 \end{pmatrix}.$$

Press 2ndMATRIX to see the menu of matrix names. Matrix names are enclosed by square brackets. Right-arrow to the EDIT menu as shown here.

```
NAMES MATH EDIT
1▮[A]
2:[B]
3:[C]
4:[D]
5:[E]
6:[F]
7↓[G]
```

Press ENTER to start the edit. Any matrix not been previously defined has the dimension 1×1. The dimension of a matrix is given as *row × column*.

```
MATRIX[A] ▮ ×1
[ 0                ]
```

Enter the dimension 2 and press ENTER. Press 2 ENTER again. The cursor is now in the matrix and ready to enter the matrix values. Initially matrix values are set to zero.

```
MATRIX[A] 2 ×2
[ 0       0       ]
[ 0       0       ]



1,1=0
```

Matrix values are most easily entered from left to right using the ENTER key after each entry. At the end of a row the cursor proceeds down and to the left of the new row, just like reading a newspaper. To make corrections use the arrow keys to position the cursor on any entry. Use 2ndQUIT when the matrix is entered correctly or use 2ndMATRIX EDIT to start editing another matrix.

Repeat all the previous steps to enter B. Matrix [B] is shown fully entered. Now that the matrices are filled, we do the required matrix arithmetic.

To calculate 5A, start with a clear home screen and enter the scalar 5. (Screen not shown.) Paste [A] from the 2ndMATRIX NAMES menu. Unfortunately there are no handy matrix names listed anywhere on the keyboard keys like there are for list names.

When the calculation is ready press ENTER; the matrix answer is shown. Repeat the process for a matrix B calculation. Notice that the matrix format starts with [[and ends with]] and that there is a row on each line.

For calculating part (c) and (d) repeat the above procedure. For each matrix name you must return to the 2ndMATRIX menu.

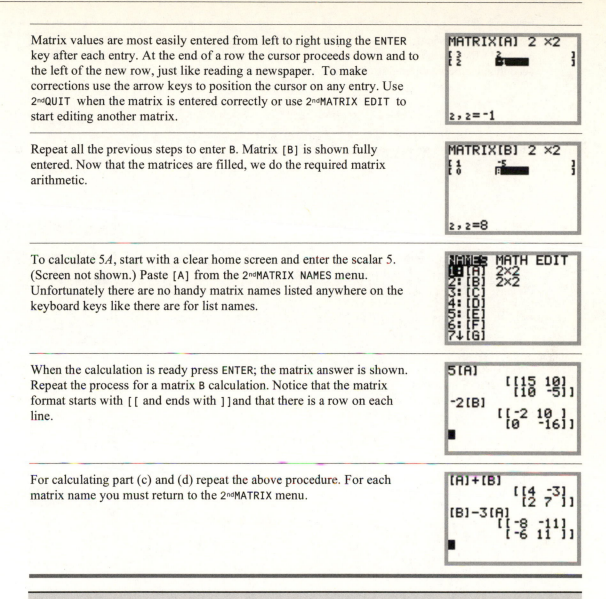

Tip: When you enter the 2ndMATRIX menu you are in the NAMES menu. To edit [B] you must arrow to EDIT before making the [B] matrix choice. If you first arrow down to [B] and then arrow to EDIT, the matrix selection resets to [A] and you have to arrow down to [B] again.

Matrix multiplication

[480] Computer graphics often require the rotation of an object about a fixed point. As an example we show how a pixel, thought of as a position vector, is rotated about the origin by a given angle. If the pixel coordinates are (2, 3) and the rotation is 30 degrees, then the new pixel coordinates are given by a matrix multiplication.

$$R\vec{P}_0 = \begin{pmatrix} \cos\theta & -\sin\theta \\ \sin\theta & \cos\theta \end{pmatrix} \cdot \begin{bmatrix} x_0 \\ y_0 \end{bmatrix}$$

Enter the rotation matrix [A] as a 2×2 matrix. The first entry was cos(30°) but it was automatically calculated to its decimal value. The second value being entered is also immediately calculated.

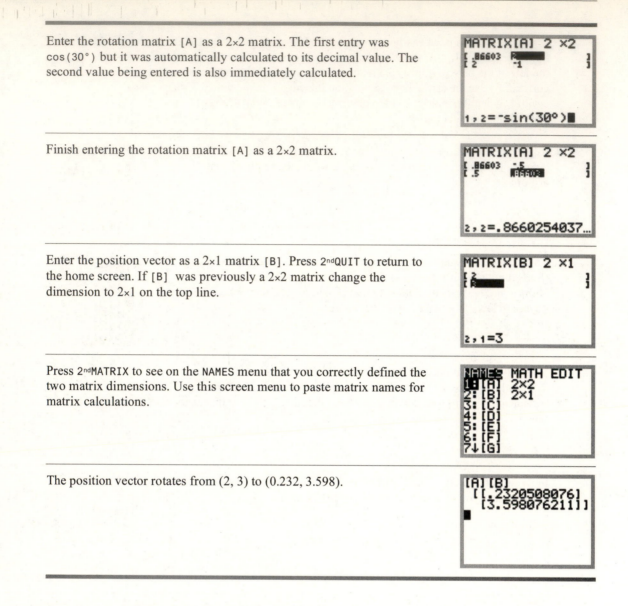

```
MATRIX[A] 2 ×2
[ .86603  ▓▓▓▓▓▓▓  ]
[ 2       -1       ]



1,2=-sin(30°)█
```

Finish entering the rotation matrix [A] as a 2×2 matrix.

```
MATRIX[A] 2 ×2
[ .86603  -.5      ]
[ .5      ▓▓▓▓▓    ]



2,2=.8660254037…
```

Enter the position vector as a 2×1 matrix [B]. Press 2ndQUIT to return to the home screen. If [B] was previously a 2×2 matrix change the dimension to 2×1 on the top line.

```
MATRIX[B] 2 ×1
[ 2    ]
[ ▓▓▓▓ ]



2,1=3
```

Press 2ndMATRIX to see on the NAMES menu that you correctly defined the two matrix dimensions. Use this screen menu to paste matrix names for matrix calculations.

```
NAMES MATH EDIT
1:[A]   2×2
2:[B]   2×1
3:[C]
4:[D]
5:[E]
6:[F]
7↓[G]
```

The position vector rotates from (2, 3) to (0.232, 3.598).

```
[A] [B]
[[.2320508076]
 [3.598076211]]
█
```

Solving linear systems of equations

We show how to use matrices to solve a system of simultaneous linear equations. This example is from the Tools section in Chapter 1 of the text. Commonly, a system of linear equations has n equations and n unknown variables. We show an example for $n = 2$, but the same technique can be used for larger values of n.

[56] Solve the system $\begin{cases} y + \dfrac{x}{2} = 3 \\ 2(x + y) = 1 - y \end{cases}$

First rewrite the two equations so that the unknowns are all on the left and the constant terms are all on the right,

$$\begin{cases} \dfrac{1}{2}x + y = 3 \\ 2x + 3y = 1 \end{cases}$$

The coefficients of the unknowns are written as one matrix, *A*, and the constants are written as a second matrix, *B*. The simultaneous solution of the two equations is found by the simple matrix arithmetic $A^{-1}*B$.

Enter the equation's coefficients in [A].

```
MATRIX[A] 2 ×2
[ 5      1      ]
[ 2      3      ]


2,2=3
```

Repeat the matrix definition process to enter the 2×1 matrix [B]. This matrix is a column matrix with the constants from the right-hand side of the system of equations. Use 2ndQUIT when the matrix is entered correctly.

```
MATRIX[B] 2 ×1
[ 3       ]
[ 1       ]


2,1=1
```

Press 2ndMATRIX and to see that you have defined the two matrices because they now show their dimensions on the NAMES menu. Use this screen menu to paste matrix names, like [A], onto the home screen for matrix calculations.

```
NAMES MATH EDIT
1:[A]  2×2
2:[B]  2×1
3:[C]
4:[D]
5:[E]
6:[F]
7↓[G]
```

Enter the calculation shown. Both [A] and [B] must be pasted from the 2ndMATRIX NAMES menu, and ⁻¹ is pasted by using the x^{-1} key. Read the answer from the matrix shown: *x* = -16 and *y* = 11.

```
[A]⁻¹[B]
           [[-16]
            [11 ]]
```

Tip: Do not enter a matrix name on the keypad. If you type 2nd[ALPHA A 2nd], the screen shows [A] but it causes a syntax error. Use 2ndMATRIX NAMES and paste a matrix name.

Tip: It may have occurred to you that an *n* × 1 or 1 × *n* matrix could be used to represent vectors. Matrices that are just one column or just one row are often referred to as vectors.

Notes:

CHAPTER ELEVEN

SEQUENCES AND SERIES

11.1 Sequences

A sequence is a list of numbers. In an arithmetic sequence, the terms increase (or decrease) by adding a fixed amount. In a geometric sequence, each term is a constant multiple of the preceding term.

Arithmetic sequence

[489]   Write a formula for the sequence of annual odometer readings for a car that is driven 8000 miles per year and had gone 15,000 miles when bought. Show the first seven years.

In Section 6.2, we introduced the seq(command in the 2nd LIST OPS menu. We use this command again to generate a sequence.

$$\text{seq}(expression, variable, begin, end, increment).$$

If the *increment* is 1 it can be omitted and 1 will be used as a default.

```
NAMES OPS MATH
1:SortA(
2:SortD(
3:dim(
4:Fill(
5:seq(
6:cumSum(
7↓ΔList(
```

Complete the command as shown to create a sequence of odometer readings. Since seven years from the date of purchase is the start of the eighth year we use 8 for the *end* value. In general using X as the variable saves keystrokes, but we used N out of respect for tradition.

```
seq(15000+(N-1)*
8000,N,1,8)
{15000 23000 31...
■
```

Use the right arrow to scroll the list and find the final mileage is 71,000 at the start of year eight.

```
seq(15000+(N-1)*
8000,N,1,8)
...00 63000 71000}
■
```

Geometric sequence

[491]   Write a formula for a salary sequence that starts at $40,000 and increases by 5% each year. Show your salary after ten years.

Since we are dealing with dollars, we change decimal rounding to 2 in the MODE.

```
NORMAL  SCI  ENG
FLOAT  01[2]3456789
RADIAN  DEGREE
FUNC  PAR  POL  SEQ
CONNECTED  DOT
SEQUENTIAL  SIMUL
REAL  a+bi  re^θi
FULL  HORIZ  G-T
SET CLOCK 01/28/07 11:06AM
```

Like the arithmetic sequence we use the seq(command to create the geometric sequence. Since after ten years from hire is the start of the eleventh year we use 11 for the *end* value.

```
seq(40000(1.05)^
(N-1),N,1,11)
{40000.00 42000...
```

Use the right arrow to scroll the list and find the salary is $65,155.79 at the start of year eleven.

```
seq(40000(1.05)^
(N-1),N,1,11)
...53.13 65155.79}
```

Other ways to create sequences

Using a function and table

Use the Y= screen to define the continuous function version of the geometric sequence.

```
Plot1 Plot2 Plot3
\Y1■40000(1.05)^
(X-1)
\Y2=■
\Y3=
\Y4=
\Y5=
\Y6=
```

Use 2nd TBLSET to start the table at 1 and increment by 1.

```
TABLE SETUP
 TblStart=1
 ΔTbl=1
Indpnt: Auto Ask
Depend: Auto Ask
```

Press 2nd TABLE to see the table. Use the arrow keys to find the desired value. Notice that two decimal display is still set in MODE, as seen in the X values. However the column width requires rounding for the Y1 values.

```
X       Y1
7.00    53604
8.00    56284
9.00    59098
10.00   62053
11.00   65156
12.00   68414
13.00   71834
Y1=65155.7850711
```

Tip: In a table, a highlighted cell's value is shown on the bottom line. It always shows full decimal accuracy, regardless of the MODE decimal setting.

Using the sequence function mode

Arithmetic sequence values are easily found by using the Seq option in the MODE menu.

```
NORMAL  SCI  ENG
FLOAT  0123456789
RADIAN  DEGREE
FUNC  PAR  POL  SEQ
CONNECTED  DOT
SEQUENTIAL  SIMUL
REAL  a+bi  re^θi
FULL  HORIZ  G-T
SET CLOCK 01/28/07 11:15AM
```

Now the Y= menu allow the entry of a sequence function $u(n)$. The u shown on the screen is not ALPHA U, but the 2nd u found above the 7 key. Also, remember that the variable key $\boxed{X,T,\theta,n}$ now is the n that is needed in the $u(n)$ definition.

```
Plot1 Plot2 Plot3
nMin=1
\u(n)◼u(n-1)+800
0
 u(nMin)◼{15000}
\v(n)=◼
 v(nMin)=
\w(n)=
```

Press 2nd TABLE to see the table. This shows that we start at 15000 and add 8000 per year. Use the arrow keys to show $n=8$ and other values.

n	u(n)
1	15000
2	23000
3	31000
4	39000
5	47000
6	55000
7	63000

n=1

11.2 Arithmetic Series

[494]   Lefkos the landscaper stacks timbers to make an upward-sloping terrace. He starts with one timber, then adds a stack of two, then a stack of three, and so on to build the terrace. How many timbers does he need if there are six stacks in the terrace?

Reset MODE to default settings. With 2nd LIST OPS paste the seq(command and complete as shown to create a sequence of the number of timbers in each stack.

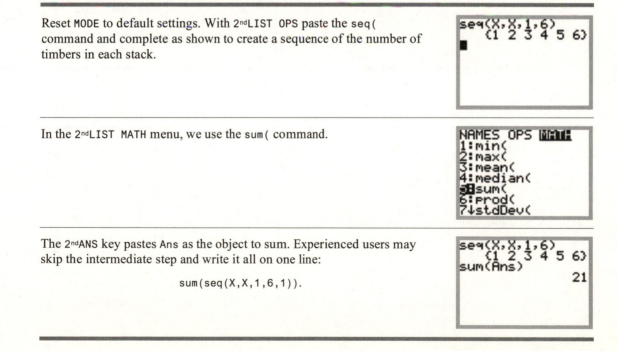

```
seq(X,X,1,6)
      {1 2 3 4 5 6}
■
```

In the 2nd LIST MATH menu, we use the sum(command.

```
NAMES OPS MATH
1:min(
2:max(
3:mean(
4:median(
5◼sum(
6:prod(
7↓stdDev(
```

The 2nd ANS key pastes Ans as the object to sum. Experienced users may skip the intermediate step and write it all on one line:

$$\text{sum(seq(X,X,1,6,1))}.$$

```
seq(X,X,1,6)
      {1 2 3 4 5 6}
sum(Ans)
                 21
```

Sum Notation

It is useful to be able to translate back and forth between the seq command and sigma notation. The previous seq(X,X,1,6) command has sigma notation: $\sum_{x=1}^{6} x$.

[497] $\sum_{i=1}^{20} (2i - 1)$ is the sigma notation for the first 20 positive odd numbers. Find the sum.

The translation from the sigma notation to the seq command is shown. The increment is not specified in sigma notation; but it is always 1.

```
sum(seq(2I-1,I,1
,20))
                 400
■
```

Tip: Any letter can be used for the variable. In the previous example, we followed the notation of the stated problem, but it requires that we use the ALPHA key. This extra keystroke is avoided by using the variable X, as much as possible.

Falling Objects

[497] An application of arithmetic sequences is finding the distance traveled by a falling object. The distance traveled per second by a falling object is given by the sequence {16, 48, 80, ...}. The linear function $y = 32x - 16$ can generate these values. Find the total distance fallen for each second.

First, we enter the sequence to check its correctness. Second, we use the cumSum command in the CATALOG (or 2ndLIST OPS menu) to generate the cumulative sums.

```
seq(32X-16,X,1,3
)
           {16 48 80}
cumSum(Ans)
          {16 64 144}
■
```

11.3 Finite Geometric Series

A finite geometric series is a sum of the form: $S_n = a + ax + ax^2 + \cdots + ax^{n-1}$.

IRAs

[501] A common application is an Individual Retirement Annuity (IRA). The balance of an IRA is a geometric series where a is the annual deposit and x is the growth rate. For example, depositing $2000 per year and getting a 6% return, we can find the balance in 5 years, right after a payment has been made.

$$S_6 = 2000 + 2000(1.06) + 2000(1.06)^2 + \cdots + 2000(1.06)^5 = \sum_{n=0}^{5} 2000(1.06)^n.$$

Once a geometric series is translated to sigma notation, the sum is easily found. The time spent changing to sigma notation is usually faster than explicitly entering the term-by-term sum. We find the sequence using `seq(` in the 2ndLIST OPS menu then we sum the sequence using `sum(` in the or 2ndLIST MATH menu.

```
seq(2000(1.06)^X
,X,0,5)
{2000 2120 2247...
sum(Ans)
       13950.63708
```

A more efficient approach is to find the sum using a formula. A geometric series is calculated directly from the formula:

$$S_n = a + ar + ar^2 + \cdots + ar^{n-1} = \frac{a(1-r^n)}{1-r}.$$

```
,X,0,5)
{2000 2120 2247...
sum(Ans)
       13950.63708
(2000(1-1.06^6))
/(1-1.06)
       13950.63708
■
```

Drug levels

[503]   A patient is given a 20-mg injection. Each day the patient's body metabolizes 50% of the drug present. Write a geometric series that gives the drug level in the body after the tenth injection.

The geometric series is

$$Q_{10} = 20 + 20\left(\tfrac{1}{2}\right) + 20\left(\tfrac{1}{2}\right)^2 + \cdots + 20\left(\tfrac{1}{2}\right)^9 = \sum_{x=0}^{9} 20\left(\tfrac{1}{2}\right)^x.$$

We find the sum in two ways. First with `sum(seq(…))` and then directly using the sum formula.

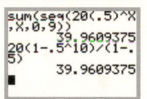

```
sum(seq(20(.5)^X
,X,0,9))
       39.9609375
20(1-.5^10)/(1-.
5)
       39.9609375
■
```

> *Tip:*   While the nested `sum(seq(…))` command is efficient, it does not show the sequence that is being summed. It is prudent to generate the list first, check the list for correctness, and then sum the list using `sum(Ans)`.

11.4 Infinite Geometric Series

We just found the finite sum of a geometric series by means of summing the terms and then by using a formula. For an infinite geometric series, summing all the terms is impossible and we must use a formula.

[506]   We concluded the previous section with the example of a patient given a daily 20-mg injection and metabolizing 50% of the drug present. Write a geometric series that gives the drug level in the body after many injections.

As the treatment goes on for more and more days, the level gets closer and closer to 40. After 100 injections, the sum is so close to 40 that it exceeds the calculator accuracy and it rounds up to 40.

Tip: The last calculation takes several seconds during which time a busy indicator scrolls in the upper right corner of the screen. If a command such as sum(seq(…)) is taking too long to calculate, press ON to break.

The infinite sum of the geometric series is calculated directly from the formula

$$S = \sum_{n=1}^{\infty} ax^n = \frac{a}{1-x} = \frac{20}{1-(0.5)} = 40.$$

Using the sum formula in this case is clearly more efficient.

The Present Value

[508]   Alex Rodriguez signed a contract with the Texas Rangers that had a signing bonus of $2 million per year for five years. The present value (using a 5% return rate) of his contract, is calculated as a geometric series:

$$Q_5 = 2 + 2\left(\tfrac{1}{1.05}\right) + 2\left(\tfrac{1}{1.05}\right)^2 + 2\left(\tfrac{1}{1.05}\right)^3 + 2\left(\tfrac{1}{1.05}\right)^4 = \sum_{n=0}^{4} 2\left(\tfrac{1}{1.05}\right)^n.$$

Since the sum has only 5 terms we could directly enter it. But since we have the sigma-notation formula we use that to find the sum. Either way we see that about $9 million is enough for the team owner to deposit to be able to meet all 5 payments.

The present value of paying Rodriguez $2 million *forever*, is algebraically derived as $S = 2/(1-1/1.05) = 42$. This value is not so easy to derive by using a large sum like we did in the previous drug example. In this case, the sum of 100 terms takes six times longer and is still not very close to the limit value.

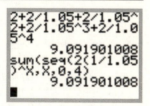

Tip: Don't count on finding an infinite sum by choosing a large end value in a sum(seq(…)) command.

CHAPTER TWELVE

PARAMETRIC EQUATIONS AND CONIC SECTIONS

12.1 Parametric Equations

We have seen that the MODE menu lists the function options Func Par Pol Seq, and we have used all but Par so far. Par is the parametric function option. The parametric way of defining a function is by using a variable t and defining the x-y position in terms of t. Thus, the Y= menu changes to define function pairs like X_{1T} and Y_{1T}, the variable key $\boxed{X,T,\theta,n}$ pastes the variable T, and the WINDOW screen begins with T-settings.

Parametric Graphing

[518]   Graph the path of a robot that begins at the point (1,0) and follows a path given by the equations:

$$x = \cos t, \text{ and } y = \sin t, \text{ where } t \text{ is in minutes, } 0 \le t \le 6.$$

We start in the Par mode. We then take the standard approach to graph a function: define the function, set the window, and graph.

We use Y= to define a single graph using a pair of functions, one for x and one for y. The variable in parametric mode is T.

We use ZDecimal to set the viewing window. (Not shown.) However, the problem stipulates $0 \le t \le 6$. We reset T values after we are satisfied with the X and Y window settings.

Trace the path to the end. It is circular but, because the *t* interval does not extend to 2π, we see that the robot stopped just short of making it back to the start.

> *Tip:* The previous example can be altered by changing Tmax to other values. Also, by changing X₁ₜ=cos(2T) and Y₁ₜ=sin(2T), the robot stays on the same path but go twice as fast.

Archimedean Spiral Revisited

[521]   In Section 7.5 a polar graph of the Archimedean spiral was shown. The graph of this spiral has parametric equations:

$$x = t \cos t, \, y = t \sin t, \text{ where } 0 \le t \le 2\pi.$$

After defining the functions, the graph is drawn in a ZStandard window. The ZStandard automatically sets T values as $0 \le t \le 2\pi$. The other ZOOM settings do not reset the T values from their current state.

> *Tip:* In parametric mode, the usual graph styles are available, but shading is not allowed.

A Parametric Walk Around the Block

[523]   Graph the motion of a particle starting at the origin and "walking" around a unit square in 4 seconds.

For this example, we use the logical expression restrictions introduced in Section 2.3 to make some rather hard to read definitions for our functions. The trick is to define the function values over the four intervals: $0 \le t < 1$, $1 \le t < 2$, $2 \le t < 3$, $3 \le t < 4$. In the definition shown below, the four intervals are written in TI notation as: (0≤T)(T<1), (1≤T)(T<2), (2≤T)(T<3), (3≤T)(T<4).

> *Tip:* The expression (0≤T)(T<1) can also be written (0≤T and T<1); this has not been used because it is longer and the definitions would not fit on a single screen.

In this definition, the first term in the tricky sum that defines X₁ₜ is T(0≤T)(T<1). This means that the function is T for $0 \le T < 1$, but 0 otherwise. The logical expressions control which formula is being used. For the graph style, we specified thick.

After entering the definitions, we use WINDOW to set Tmin=0, Tmax=4, and Tstep=.1. Graph with ZDecimal. Press TRACE to produce the screen shown. Notice that both definitions show across the top, but they are truncated, because they are so long.

> *Tip:*   The parametric equation name Y₁ₜ when shown as a label at the top of a graph window
>   appears as Y1ₜ. The 1 is raised.

> *Tip:*   Many beautiful curves are drawn using the parametric mode. Try the Lissajous figure:
> $x = \cos(3t)$, $y = \sin(5t)$ with a ZStandard and a ZoomIn.

12.2 Implicit Defined Curves and Circles

Curves created by slicing (sectioning) a cone, are appropriately called conic sections. We have already learned to graph one conic, the parabola. We demonstrate how any function defined in the Func graph mode is graphed using the parametric mode (Par). In this section, all angles are measured in radians, so be sure to have the angle mode set to Radian.

Parabola

To graph $y = f(x)$ in the parametric mode, use $x = t$, and $y = f(t)$. For example, the parabola has the equation $y = f(x) = x^2$, so we define X₁ₜ=T and Y₁ₜ=T².

After entering the above parametric equations, we set the window as shown. First press ZDecimal and ignore the graph. Next press WINDOW and enter the T values to be the same as the X values as shown.

Pressing TRACE allows us to see the equation definitions. Notice that the trace starts at the Tmin value, which in this case is a point off the screen. In contrast, TRACE starts at the screen center in the Func mode.

In Section 8.2, we introduced DrawInv to graph the inverse of a function. In the parametric mode, inverses are simple to graph, and they can even be traced. Simply switch the X and Y definitions. For example switching these equations, as shown at the top of the screen, draws the inverse of the parabola.

Circles

For $r > 0$, the parametric equations of a circle of radius r centered at the point (h,k) are

$$x = h + r\cos t \text{ and } y = k + r\sin t, \ 0 \le t \le 2\pi$$

[527]   Graph the parametric equations $x = 5 + 2\cos t$ and $y = 3 + 2\sin t$.

In Par mode, enter the two equations and graph with ZStandard. This does not look like a circle! The problem is that the ZStandard is not a square setting. It does give us a good idea about the center and what might be needed to have a better window.

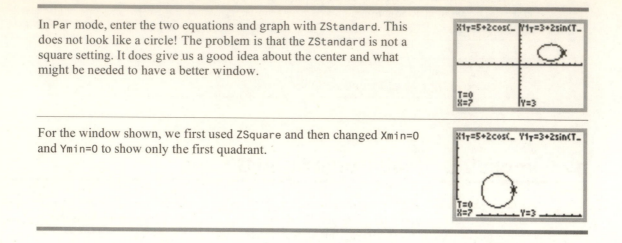

For the window shown, we first used ZSquare and then changed Xmin=0 and Ymin=0 to show only the first quadrant.

12.3 Ellipse

The parametric equations of an ellipse centered at (h, k) are:

$$x = h + a \cos t, \ y = k + b \sin t, \text{ for } 0 \le t \le 2\pi.$$

[529]   Graph the parametric equations $x = 7 + 5 \cos t$ and $y = 4 + 2 \sin t$.

In this graph we used the window settings from the previous example. These settings are square so that we are assured that this is a noncircular figure.

Tip: As in the parabola example, we can switch the major and minor axes of an ellipse by simply interchanging the equation definitions.

12.4 Hyperbola

The parametric equations of a hyperbola centered at (h, k) are:

$$x = h + a \sec t, \ y = k + b \tan t, \text{ for } 0 \le t \le 2\pi.$$

[534]   Graph the equation $\dfrac{(x-4)^2}{9} - \dfrac{(y-7)^2}{25} = 1$.

We first translate this equation to parametric form. The center coordinates are found from the numerators, $(h, k) = (4, 7)$. The values of a and b are found from the denominators as $a = \sqrt{9} = 3$ and $b = \sqrt{25} = 5$.

Enter the functions with special care since there is no secant function. $X_{1T}=4+3/\cos(T)$ and $Y_{1T}=7+5\tan(T)$. Use a ZStandard window to get a first view.

The window settings for T have been left as $0 \le t \le 2\pi$, but the X and Y window values are reset to give a better view.

This shows the characteristic figure for a hyperbolic function. You may be wondering about the X-lines in the middle of the screen. This is the connection across discontinuities that we previously encountered for rational functions. As before, use Dots mode to make the lines disappear or just think of them as asymptotes.

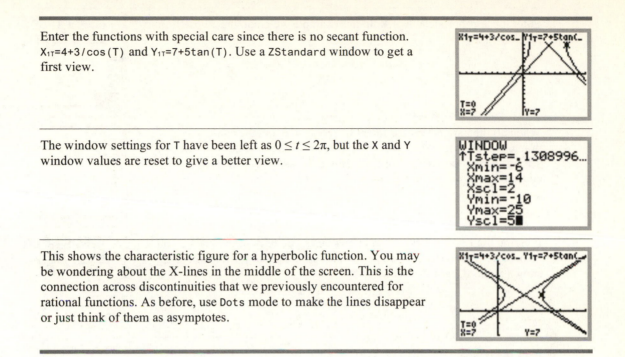

APPS on the TI-84 and TI-83 Plus

Users of the older TI-83 (non-Plus) do not have APPS capability and should skip this section. Some applications come preloaded on your TI. Many applications are available for downloading from the TI website. See *http://education.ti.com*. Select Downloads, then Download APPS. See the appendix for notes about the TI to computer connection.

Press APPS and a window similar to the one shown appears with a menu of available applications.

After selecting 6:Conics, select 2:ELLIPSE. Notice that two of the top graph row keys have been redefined with new meanings:

Y= activates INFO and GRAPH activates QUIT.

The next step is to specify the major/minor axis. Choose 1: for a horizontal major axis. Choose 2: for a vertical major axis. The ESC key (Y=) takes you back to the previous screen.

Enter the appropriate A,B,H,K. In this example we graph the ellipse from Section 12.3.

When all the parameters are entered, press GRAPH and the conic is graphed in an automatically set window. Unfortunately, the window settings cannot be seen or changed. However, TRACE can be used to see coordinates of points on the graph.

Pressing the ESC key (Y=) several times takes you back to the main menu. Choose 3:Hyperbola, 1:, and enter the parameters of the hyperbola shown at the start of this section.

Now press GRAPH or TRACE to see the graph.

12.5 Geometric Properties of Conic Functions

The focal points of conics are of interest when we investigate the reflective properties of a conic curve. We use a program to simplify the calculation of the focal points of an ellipse. A similar program is easily constructed for the other conics. The program is entered most easily on a computer using *TI-Connect™*, but it is also easily entered on the TI using PRGM NEW and PRGM EDIT. All the commands with lowercase letters are on the menus of PRGM CTL and PRGM I/O.

The name given this program is FOCALE.

```
ClrHome
Disp "X=H+A*COS(T)"
Disp "Y=K+B*SIN(T)"
Prompt H
Prompt A
Prompt K
Prompt B
ClrHome
Disp "FOCAL POINTS"
Disp "X1="
Disp "Y1="
Disp "X2="
Disp "Y2="
If A>B
Then
√(A²-B²)→C
Output(2,4,H+C)
Output(3,4,K)
Output(4,4,H-C)
```

```
Output(5,4,K)
Else
√(B²-A²)→C
Output(2,4,H)
Output(3,4,K+C)
Output(4,4,H)
Output(5,4,K-C)
End
Stop
```

The program displays the standard parametric equations of an ellipse and prompts for the parameters.

```
X=H+A*COS(T)
Y=K+B*SIN(T)
H=?█
```

We enter the parameters from the ellipse in Section 12.3.

```
X=H+A*COS(T)
Y=K+B*SIN(T)
H=?7
A=?5
K=?4
B=?2█
```

We see the display of the coordinates for the two focal points. Note that these values are not store in variables named X1, Y1. However the value of c, called the focal length, is stored in the named variable C.

```
FOCAL POINTS
X1=11.58257569
Y1=4
X2=2.417424305
Y2=4
```

12.6 Hyperbolic Functions

The hyperbolic sine and hyperbolic cosine are defined by

$$\cosh(x) = \frac{e^x + e^{-x}}{2} \quad \text{and} \quad \sinh(x) = \frac{e^x - e^{-x}}{2}.$$

Graphing the Hyperbolic Functions

[552]   Graph the cosh and sinh functions.

Use the Func graphing mode. Shown here is the cosh(in the CATALOG. Paste this to Y1 to define the function.

```
CATALOG        ▲
  Connected
  CoordOff
  CoordOn
  cos(
  cos⁻¹(
▶ cosh(
  cosh⁻¹(
```

We graph in a ZDecimal window.

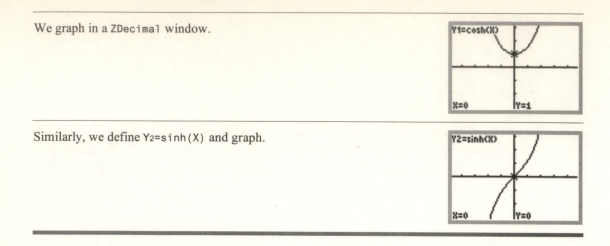

Similarly, we define Y2=sinh(X) and graph.

The Identity cosh²x - sinh²x = 1

[554]   The identity $\cosh^2 x - \sinh^2 x = 1$ can be graphed like its counterpart $\cos^2 x + \sin^2 x = 1$.

The expression $\cosh^2 x$ is written in any of the following ways: cosh(X)², (cosh(X))², or as cosh(X)^2. The left side of the identity is written as Y1 and graphed in a ZDecimal window,

Parameterizing the Hyperbola Using Hyperbolic Functions

[554]   Compare the curve parameterized by the equations

$$x = \cosh(t), \ y = \sinh(t), \text{ for } -\infty < t < \infty$$

and the hyperbola defined by $x^2 - y^2 = 1$.
First graph $x^2 - y^2 = 1$ as the parametric equations: $x = \sec t$, $y = \tan t$.

In Par mode, enter the two trigonometric functions in X1T and Y1T. Use ZDecimal and $-\pi \le t \le \pi$. We see the graph is centered at the origin and has two branches opening left and right. Think of the lines through the origin as asymptotes.

Turn off the first pair. Now enter the second pair of equations as hyperbolic functions in X2T and Y2T. Graph to see that this is the same as the rightward opening branch of the hyperbola graphed above.

Internet Addresses

The main Internet address for Texas Instruments calculators is

http://education.ti.com

While the site is full of good information, the critical part of this website is in the Download Center which is reached by selecting DOWNLOADS on the top banner of the home page. The August 2006 version is shown below

Three items of interest are shown in this upper portion of the homepage:

Guidebooks by Calculator – These are available for each calculator model.

Calculator SoftwareApplications (Apps) - A source for APPS.

Computer Software – Where to get *TI-Connect™* .

Further down the TI homepage is Program Archives, which has some classic programs that have been written. However, the most extensive archive of programs (and games) is at:

http://www.ticalc.org

Linking Calculators

The essentials of linking are presented in the *TI–Guidebook* and are not repeated here. Calculator-to-calculator connections have become less important as the Calculator-to-computer connection has become simple and more popular. But we include the following tips.

♦   The end-jack must be pushed firmly into the socket. There is a final click you can feel as it makes the proper connection.

♦   If you are experiencing difficulty connecting, turn off both calculators, check the connection, and then turn them on and try again. If available, try other cables or calculators.

♦   When selecting items, the cursor square is hardly visible when the selection arrow is on the same item. As you arrow off of the item, double-check whether or not it has been selected.

♦   If you are required to drain your calculator memory before entering an exam, use 2ndLINK Back Up to keep a copy on some other calculator. Even better, store it on your computer, which is the next topic.

Linking to a Computer

The TI-84 connects to your computer via a USB cable and the software *TI-Connect*™. The TI-83 requires a *TI-Graph Link*™ cable and the *TI-Connect*™ software. There are many advantages to using *TI-Connect*™.

♦   This is the best way to back up your work.

♦   You can download apps from the TI website.

♦   It is the preferred way to write and edit programs.

♦   You can download and transfer programs from the Internet archives.

♦   It allows you to capture the screen in a form for direct printing or use in a word processor.

Troubleshooting

Nothing shows on the screen

♦   Check the contrast.

♦   Check the ON/OFF button.

♦   Pull out the batteries and correctly reinsert the batteries.

♦   As a last resort, remove the backup battery and reinsert all the batteries. *Warning:* This erases all memory, including programs.

Nothing shows up on the graph screen except the axes

♦   Press TRACE to see if a function is defined but outside the window.

♦   There may be no functions selected.

♦   The function may be graphed along either axis and need the window reset.

♦   If there is a busy indicator (a running line) in the upper right corner of the screen, then the TI may be still calculating. Press ON if you can't wait.

♦   If there is a pause indicator (a twinkling line) in the upper right corner of the screen, then the TI is paused from a program. Press ENTER to continue.

♦   If there is a full checkerboard cursor, then you have a full memory. You need to delete something; choose some things you no longer need or copy them to your computer.

Nothing shows up in the table

You may have the Ask mode set and need to either enter *x*-values or change to Auto in the 2ndTBLSET.

Check to see that a function is selected.

I get a syntax error screen

The most common errors are

♦   Parentheses mismatch. Count and match parentheses carefully.

♦   Subtraction versus negative symbol. For example, the subtraction sign cannot be used to enter Xmin=-10.

♦   Pasting a command in the wrong place. For example, a program name must be on a fresh line.

I get an error message

This can cover the widest array of problems. Read the message carefully; it should tip you off to the kind of error you are looking for. If you have no idea what could have caused it, consult the appendix of the *TI–Guidebook* for explanations of the error messages. If that fails, go to education.ti.com and use the TI-Cares™ KnowledgeBase search found in FAQs.

I'm getting a result but it is wrong

Check for

♦   Typing direct letters in place of pasting a command. For example, SIN(X) can be entered using the ALPHA keys, but it is not the same as sin(X) using the SIN key.

♦   Parentheses mismatch. Count and match parentheses carefully.

♦   Subtraction versus negative symbol. For example, if the negation sign is used to enter X-10, the result will be −10 times the value of X and not X minus 10.

♦   Correct default settings. For example, check the MODE for the Radian/Degree angle setting.

My program won't run

Program errors are difficult to diagnose. Scroll and check your code with PROGRAM EDIT. Add temporary displays and pauses to check the progress and isolate the problem. If you created the program using *TI-Connect*™ and then transferred it to the calculator, use the right arrow key to scroll through the code to find any (invisible) spaces at the end of a line.

Notes:

Index

Notes:

Notes:

Notes:

Notes:

Notes:

Notes:

Notes:

Notes: